Advancements in Fused Deposition Modeling (FDM) Technology for Pharmaceutical and Medical Applications

Advancements in Fused Deposition Modeling (FDM) Technology for Pharmaceutical and Medical Applications

Kuldeep Vinchurkar, PhD, MPharm
Dinesh Kumar Mishra, PhD, MPharm
Pankaj Dixit, PhD, MPharm
Sheetal Mane, PhD, MPharm
Editors

AAP | APPLE ACADEMIC PRESS

Foreword

In the rapidly evolving landscape of healthcare, the fusion of technology with pharmaceutical and medical sciences has heralded a new era of innovation and progress. *Advancements in Fused Deposition Modeling (FDM) Technology for Pharmaceutical and Medical Applications* serves as a beacon illuminating the transformative journey of 3D printing technology in healthcare.

This edited volume is a testament to the collaborative efforts of esteemed editors and contributors who have meticulously curated a collection of insightful perspectives and groundbreaking research. Through its pages, we embark on a journey that traverses the intricate intersections of FDM technology with pharmaceutical and medical realms.

As we delve into the depths of this dynamic field, we witness the myriad ways in which FDM technology is reshaping healthcare delivery. From the fabrication of personalized drug-delivery systems to the creation of intricate anatomical models for surgical planning, each chapter unveils a new facet of the transformative potential inherent in FDM technology.

The pace of innovation in FDM technology demands continuous reflection, collaboration, and exploration. This volume not only captures the current state of advancements but also serves as a catalyst for future breakthroughs, inspiring dialogue and pushing the boundaries of what is possible in pharmaceutical and medical applications.

We extend my heartfelt appreciation to the editors, contributors, and all those involved in the creation of this seminal work. May *Advancements in Fused Deposition Modeling (FDM) Technology for Pharmaceutical and Medical Applications* serve as a guiding light, inspiring researchers, clinicians, and industry professionals to chart new paths toward enhancing patient care and advancing healthcare delivery.

—***Authors***

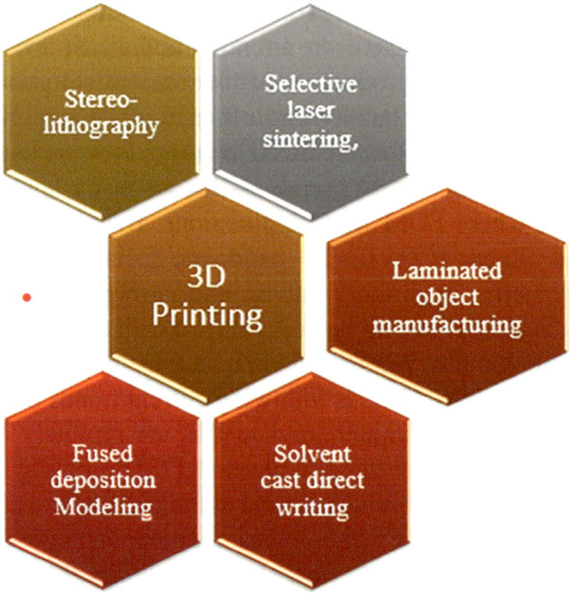

FIGURE 1.1 3D-printing in pharmaceuticals.

1.2.1 USING STEREO-LITHOGRAPHY METHOD

Stereolithography, or SLA, is a 3D printing technique that employs a liquid resin photo-curing procedure. After pouring a liquid resin mixture into a tank, a laser is used to scan the resin surface in a pee-programmed pattern to start photo-polymerization. This causes the resin to solidify through a cross-linking procedure using chemicals. SLA is widely used in many applications, from prototyping consumer products to printing living tissues. However, SLA has two major issues that need to be addressed: processing time and parts performance, particularly their thermo-mechanical properties [12].

1.2.2 USING SELECTIVE LASER SINTERING METHOD

The industrial 3D printing method known as selective laser sintering (SLS) builds up the 3D object using a powder bed. Similar to PB, SLS binds the powder particles together with a laser rather than a spray solution. The laser creates an exact pattern on the powder bed's surface during the printing process.

After the first layer of powder is applied, another layer is rolled on top. This procedure is carried out to construct the object by repeatedly adding layers of powder until it is recovered from beneath the powder bed. Compared with PB, selective laser sintering or SLS technology has a few advantages. For example, compared with PB, it enables faster output and is a solvent-free method. On the other hand, for the solvent to evaporate while using the PB technology, the printed object must be left for up to 48 h [13].

1.2.3 USING LAMINATED OBJECT MANUFACTURING (LOM) METHOD

The laminated object manufacturing (LOM) process is a method of something for creating prototypes using layers of laminated materials such as paper, polymeric films, foils, and laminated metal. Paper has inferior mechanical qualities compared with polymeric foils. The process involves joining and clamping the sheets together using adhesion and ultrasonic welding. Following the joining of the sheets, a layer is applied, and the material is cut into the final shape of the product using a laser beam or knife.

Solido, an Israeli company, uses a variation of LOM to create prototypes. They apply a glue layer to a film, which is then cut into the desired shape by a knife. An "anti-glue" layer is put on the areas where no prototype is desired, neutralizing the glue. The next layer of film is then applied and glued returning to the layer before, repeating the process until the final layer is complete. Once the final layer is finished, excess material is manually removed from the prototype [14].

1.2.4 USING SOLVENT CAST DIRECT WRITING (SC-DW) METHOD

Another way to process thermoplastics is through direct write deposition in solvent-cast 3D printing. In this method, the materials are dissolved in a solvent or solvent combination to form a highly viscous polymer solution. This solution is then extruded through a nozzle to create a filamentous extrude with a diameter ranging from 1 mm to 100 μm. The extruded material quickly dries and hardens to hold its form as the subsequent layer is applied. This process is performed at room temperature and eliminates the need for high-temperature processing of polymers. Furthermore, this method allows for the combination of materials, including other polymers and inorganic materials, to create a composite ink, provided that suitable solvents are used [15].

1.2.5 FUSED DEPOSITION MODELING (FDM)

One method for quick prototyping is FDM. It had its start in Tucson, Arizona's Advanced Ceramics Research (ACR) Company. Later, it was greatly enhanced by the Stratasys Corporation. A 3D CAD model is used as the starting point of the procedure, and computer software divides it into horizontal layers. The machine receives the filament (such as polymer or wax) through a nozzle that is computer-controlled and creates the raster of the object in that layer. After leaving the nozzle in a liquid state, the substance rapidly solidifies at room temperature. Usually, the system is housed in a heated space. Following the creation of the first layer, the new layer is extruded and the working bed is lowered to accommodate its thickness. A twin extrusion head is employed in this instance. The thickness of the layer relies on the nozzle opening, material dosage, and feeding rate. Typically, it ranges from 0.18 to 0.5 mm.

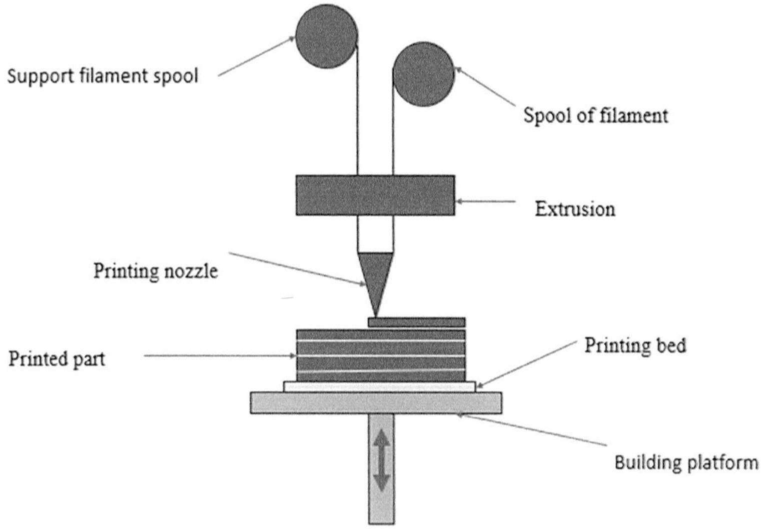

FIGURE 1.2 Schematic diagram of the FDM process.

It is possible to extrude biocompatible and biodegradable materials such as polycaprolactone (PCL) and elastomers and to produce multiple prototypes simultaneously [16]. The material is fed into the liquefier head as a filament with the help of the grooved bearing and drive gear, as depicted in Figure 1.2. Before being deposited in layers using a heated nozzle, the thermoplastic filament is heated to a glass transition temperature. Following the tool route that programming provides, the liquefier head moves through the X-Y plane. Once fabrication is completed, support material can be removed with a solvent [17].

1.3 COMPREHENDING THE FUSED DEPOSITION MODELING PROCESS PARAMETERS (FDM)

Many parameters affect FDM, as Figure 1.3 summarizes. As can be seen, the parameters are divided into two major groups: material parameters and machine parameters. The material properties of the filament used in

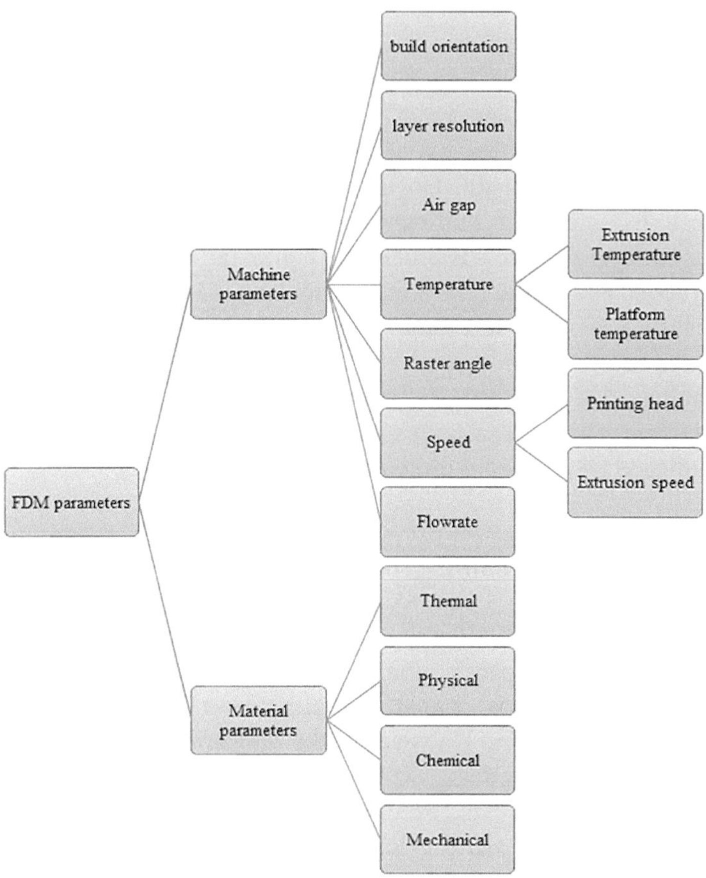

FIGURE 1.3 Different types of FDM process parameters.

the 3D printer are referred to as material parameters. On the other hand, the machine parameters are the settings that the user inputs into the slicer software when generating the G-code files for printing. The printing speed, raster angle, melt flow rate via the nozzle, air gap, layer thickness, infill density, build orientation, and temperature are a few of the machine parameters that are displayed on the other hand; material properties such as thermal and mechanical influence both the extrusion and performance of the print [18, 19]. Here, discuss some critical FDM process parameters, and their influences on the resulting parts.

1.3.1 LAYER RESOLUTION (LAYER THICKNESS)

Layer thickness, also known as layer height, refers to the thickness of each plastic layer deposited during 3D printing. The impact of layer thickness on the final product cannot be overstated. Generally, thicker layers result in faster builds, but at the cost of a poorer surface finish and potentially lower mechanical strength. On the other hand, thinner layers improve surface finish and detail but increase printing time. Therefore, selecting an optimal layer thickness is crucial for achieving desired results in terms of both speed and quality. As a result, choosing the ideal layer thickness is essential to getting the intended outcomes in terms of speed and quality. It requires carefully considering some factors, such as the type of printer, the material, and the purpose for which it is used. Achieving the ideal layer thickness that provides an equal amount of speed and quality can be accomplished by balancing these factors [20].

1.3.2 EXTRUSION TEMPERATURE

To guarantee the quality of the finished product, the thermoplastic material must be heated to a specific temperature before extrusion. Maintaining the ideal temperature is crucial to facilitate proper material flow and adhesion between layers. However, if the temperature is too high, it can result in stringing and oozing, which can compromise the integrity of the finished product. However, excessively low temperatures can result in inadequate layers and increase brittleness. Therefore, it is imperative to carefully monitor and control the temperature during the extrusion process to achieve optimal results [21].

1.3.3 PRINT SPEED

When it comes to 3D printing, how quickly the print head moves while extruding material is a crucial factor that affects both build time and print quality. Faster print speeds can certainly reduce build time, which is desirable in many cases. However, this may come at the cost of compromising print quality due to insufficient cooling or poor adhesion. On the other hand, slower speeds tend to improve accuracy and surface finish, but they also prolong the overall printing process. Thus, to get the best outcomes from 3D printing projects, it is crucial to strike the correct balance between speed and quality. By carefully adjusting the print speed according to the specific requirements of each project, one can ensure that both efficiency and precision are maximized without sacrificing either aspect [22].

1.3.4 INFILL DENSITY AND PATTERN

Pattern and infill density are two crucial aspects of 3D printing that need to be taken into account. When creating a 3D print, the amount of material used to fill the inside is an important factor to consider. This is known as infill density and can greatly impact the strength and durability of the finished product, and the pattern dictates how this material is arranged. It is crucial to remember that a printed part's strength and mechanical qualities can be greatly impacted by both infill density and pattern. Increasing the infill density will generally result in a stronger part, as there is more material present to support the structure. However, higher infill densities also require more material and time for printing, which can increase costs and production time. Under certain loading circumstances, the infill pattern can also have an impact on a printed part's mechanical characteristics. Certain patterns might be more appropriate for tension or bending stresses, whereas others might offer superior support in compression.

Overall, it is important to carefully consider both infill density and pattern when designing and printing parts using 3D technology. By understanding these concepts, designers can optimize their prints for strength, functionality, and cost-effectiveness [23].

1.3.5 NOZZLE DIAMETER

An important consideration in 3D printing is the printer nozzle diameter, which has a direct impact on print quality and speed. The nozzle is responsible for extruding the material onto the printing surface, and its size determines how much material can be extruded at once. A larger nozzle has a higher capacity to extrude more material, which can significantly increase the printing speed. However, this comes at a cost as larger nozzles may reduce the detail level of the print due to their lower precision. On the other hand, smaller nozzles are ideal for producing intricate designs with high levels of detail but at a slower pace since they have a limited capacity to extrude material. Hence, choosing the right nozzle diameter involves striking a balance between print speed and detail level to achieve optimal results in 3D printing [24].

1.3.6 BED TEMPERATURE

The temperature of the build platform is one factor that is very important in 3D printing. Better adhesion between the printed product and the surface it is being produced on is encouraged by the heated build platform. Consequently, this contributes to the ultimate result being more exact and precise. Additionally, it is important to remember that the build platform's temperature also affects the rate at which the bottom layer cools down. By adjusting this parameter, users can optimize their 3D printing process for their specific needs and create high-quality products with ease [25].

1.3.7 RETRACTION SETTINGS

Retraction is a technique used in 3D printing where the filament is pulled back during movement when it is not being extruded. This process helps prevent stringing or oozing, which can negatively impact the overall quality of a print. It is crucial to have proper retraction settings in place to ensure clean prints without any unwanted strings between different parts of the print. Without appropriate retraction, the print could end up with unsightly defects that detract from its overall appearance and functionality. Therefore, it is essential to pay close attention to retraction settings and adjust them as needed to achieve optimal results [26].

To achieve the desired outcomes in print quality, efficiency, and material properties, the interdependent parameters must be optimized together. This requires a thorough understanding of how these parameters work together and the impact they have on the result. Production engineers often rely on a trial-and-error approach to finding the optimal settings for their specific application, material, and printer. However, it can be an expensive and time-consuming procedure. The use of the design of experiments (DoE) approaches can help to expedite the optimization process. By using DoE techniques, production engineers can systematically test different combinations of parameters to identify the optimal settings that will yield the desired outcomes. This approach not only saves time and resources but also ensures that the print quality, efficiency, and material properties are optimized to meet or exceed expectations [27].

1.4 APPLICATION

Compared with 3D printing, the production flexibility and process capabilities of the traditional techniques of generating medications are severely limited. Recent years have seen a noticeable increase in the impact of AM due to the flow in demand for customized medications and medical equipment [28]. The ability to create precisely personalized dosage forms, tissue structures, and illness mimics has made 3D printing an essential and revolutionary technology. It is beneficial for creating customized products in a variety of industries due to its versatility and ability to create complex designs, including implants and medical equipment. Additionally, FDM technology is more and more in the medical technology to create molds for casting implants and medical equipment. The typical casting of investment typically involves the use of metallic molds and sacrificial patterns like wax to achieve intricate shapes for implants. By utilizing 3D-printed molds, the necessity for sacrificial materials is eliminated, thereby reducing costs, time, and material waste [29]. The FDM technology has emerged as one of the most progressive methods for producing prototypes and rapid tooling of complex products in low and medium batches [30].

In the pharmaceutical field, the integration of FDM with hot melt extrusion (HME) and finely tuned formulation compositions has emerged as a promising avenue for crafting pharmaceutical tablets and implants necessitating customizable drug release profiles.

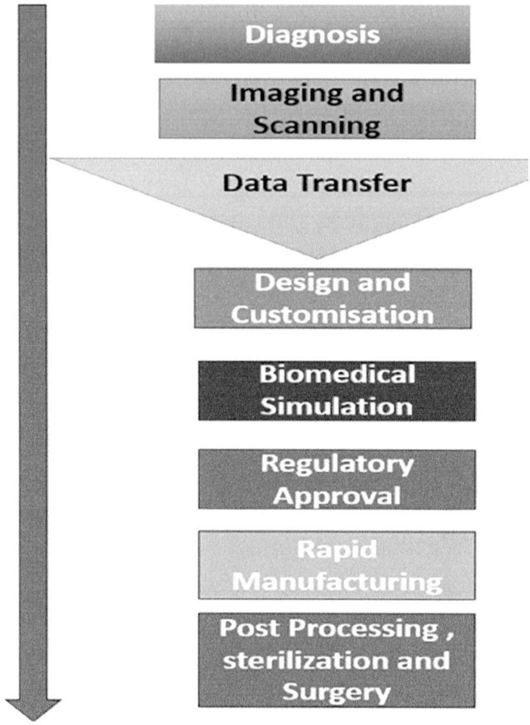

FIGURE 1.4 Flowchart of additive manufacturing in medical applications.

1.4.1 PATIENT-CENTRIC THERAPY

Over the last decade, there has been a significant emphasis on developing drug products that prioritize the needs and experiences of patients. Patient-centric therapy revolves around tailoring medical treatments and health-care services to meet the unique needs, preferences, and circumstances of individual patients [31].

1.4.2 CUSTOMIZED AND PERSONALIZED DRUG-DELIVERY SYSTEMS

FDM contributes to the development of custom-made drug doses designed to suit the unique requirements of each patient. By integrating specific information about the patient, like age, weight, and medical background, pharmaceutical preparations can be tailored to enhance treatment effectiveness and reduce adverse reactions. FDM 3D printing technology is feasible for fabricating personalized tablets containing drug doses tailored to individual patients or tablets with specific drug-release profiles [32].

1.4.3 PRECISION MEDICINE

Another area where the FDM technique holds promise is in the domain of precision medicine [33]. Currently, the adjustment of the suitable dosage for individual patients is predominantly achieved through liquid formulations, such as injectables or oral solutions, with oral solid forms being rarely utilized. However, liquid formulations pose challenges due to their inherent instability, the necessity for additional materials such as syringes and dosing pipettes, and sometimes requiring nursing intervention for administration [34]. Liquid dosage forms are more likely to be tampered with and to have dosing errors than other forms of medication. Therefore, the FDM technique presents an opportunity to provide patients with customized oral solid formulations that are safer and easier to administer [35].

1.4.4 PEDIATRICS

The capacity to create medications with personalized dosage strengths, flavors, shapes, and sizes can offer numerous advantages for pediatric populations, for whom conventional mass-produced formulations may not be ideal due to factors such as poor taste or inappropriate dosages [36]. Numerous studies have concentrated on developing child-focused formulations by using 3D printing because there is limited research on suitable solid dosage forms for children. Numerous innovative dosage form designs such as mini-tablets, orodispersible films, orodispersible tablets, and capsules have showcased this capability [37, 38]. It is believed that such inventive designs will exhibit high acceptability in children, potentially due to more attractive colors and dimensions [39, 40].

1.4.5 GERIATRIC

The elderly care sector stands to benefit from the application of FDM technology [41]. Generally, the elderly population exhibits alterations in organs responsible for metabolizing medications, such as the kidneys, and heightened sensitivity to certain compounds, including psychotropic and anticholinergic agents. Consequently, there is often a need to adjust medications to lower doses [42]. Additionally, elderly patients may be prescribed multiple medications daily, sometimes exceeding 10 drugs, leading to complex medication management and reduced adherence. FDM technology plays a crucial role in the geriatric population as printed medications feature sustained-release kinetics, which is typically associated with fewer side effects, reduced frequency of drug administration, and combination of two or more active pharmaceutical ingredients (APIs) in a single therapeutic unit [43].

1.4.6 PATIENT-SPECIFIC IMPLANTS AND PROSTHETICS

The FDM system supports the manufacture of patient-specific implants, prosthetics, and orthotics with precise anatomical fit and functionality. Medical imaging data, such as CT scans or MRI images, can be utilized to aid in the diagnosis and treatment of various medical conditions, FDM can produce implants that closely mirror the patient's framework. This personalized approach enhances comfort, function, and long-term outcomes for the patient [44].

1.4.7 BIO-ROBOTICS

In recent years, there has been a notable surge in interest in bio-inspired hybrid devices capable of replicating various biological functions [45]. These bio-robots are built with hydro-gels or polymer elastomer-based synthetic scaffolds that support soft biological elements such as proteins, live cells, or tissues. They are more flexible than typical robots, enabling a variety of movements like walking and swimming as well as enabling contact with the environment. Rotary machines, which are primarily involved in converting chemical energy from ATP hydrolysis into mechanical work, stand out as particularly inspiring among these robots [46].

1.4.8 WOUND DRESSING

The development involved creating customized antimicrobial dressings for wounds using polycaprolactone, or PCL infused with elements of copper, zinc, and silver. This process included producing filaments uniformly loaded with metals via HME, followed by 3D printing models of the nose and ears. These dressings demonstrated the ability to release different metals over an extended period and to counteract bacteria. Their adaptive design tailored to individual anatomy made them more cost-effective compared with traditional flat dressings. Additionally, a hybrid scaffold combining poly (ethylene glycol) (PEG) and homogenized pericardium matrix was engineered via 3D printing to enhance wound healing in vascular grafts. These scaffolds are intended to facilitate the replacement of damaged vessel walls following surgical reconstruction [47].

1.4.9 DENTISTRY

FDM technology is widely applied in bio-medicine, encompassing the creation of dental models for dentistry and associated items like denture flasks, bite guards, mouth guards, and oral drug delivery devices [48]. Polyether Ether Ketone (PEEK) has gained prominence as a frequently suggested material in oral and craniomaxillofacial surgery, enabling its utilization in crafting dental implants and manufacturing dental crowns [49].

1.4.10 NOVEL CORONAVIRUS (COVID-19) PANDEMIC

During the COVID-19 pandemic, FDM technology was utilized to produce face masks manufactured using 3D technology that uses functionalized graphene-coated air filters. Additionally, FDM was employed to manufacture ventilators, respirator systems, respiratory valves, and nasopharyngeal swabs, aiding in COVID-19 diagnosis and treatment efforts [50].

1.4.11 DRUG DEVELOPMENT RESEARCH

The research field stands to gain significant advantages from FDM-3D printing, particularly in its potential application for drug development spanning from initial phases to human clinical trials. FDM allows researchers to create innovative drug delivery systems with precise control over the kinetics of drug release and dosage forms. By 3D printing drug-loaded matrices, micro-particles, or implants, researchers can explore various formulations and optimize drug delivery strategies for enhanced efficacy, reduced side effects, and improved patient compliance [51].

1.4.12 TISSUE ENGINEERING AND DRUG TESTING

FDM enables the fabrication of 3D tissue scaffolds and organ-on-a-chip models for drug testing and disease modeling purposes [52]. By printing bio-mimetic tissue structures with precise geometries and compositions, researchers can simulate physiological conditions, study drug responses *in vitro*, and predict drug efficacy and safety before clinical trials [53].

1.4.13 PHARMACOKINETIC MODELING AND SIMULATION

FDM facilitates the fabrication of anatomically accurate pharmacokinetic models and drug diffusion simulations for pharmacokinetic studies. By 3D printing patient-specific anatomical structures and drug distribution models, researchers can simulate drug absorption, distribution, metabolism, and excretion processes, providing valuable insights into drug pharmacokinetic and tissue targeting [54].

1.4.14 TASTE MASKING

FDM allows for the fabrication of customized drug dosage forms with controlled release properties, which can help mask unpleasant tastes. By incorporating taste-masking agents or encapsulating drugs within taste-neutral materials during the printing process [55], FDM can produce oral dosage forms that minimize the perception of bitterness or other undesirable tastes upon administration. FDM enables the creation of layered structures with different materials, textures, or flavors. FDM designs multilayered tablets or capsules with an outer shell containing flavoring agents or sweeteners to mask the taste of the drug core. This layering approach provides a barrier between the drug and the taste buds, improving palatability and patient acceptance [56].

TABLE 1.1 Examples of 3D-Printed Medicine by FDM.

Manufacturing Method	Dosage Form	API	Excipients	Effects	References
FDM 3D printers	Intragastric floating sustained-release tablets	Domperidone (DOM)	Hydroxypropyl cellulose (HPC)	Produce hollow tablets, allowing them to float because they have less density.	[57]
Dual extruder FDM 3D printers	Gastric-resistant delayed-release tablets with a shell core.	–	PVP, methacrylic acid polymer	Delayed release solid dosage forms	[58]
FDM 3D printing	Modified-release tablets	Paracetamol	Hypromellose acetate succinate (HPMCAS) (grades LG, MG, and HG)	Produce delayed-release tablets that do not require an external enteric coating.	[29]
Dual extruder FDM 3D printer	Controlled-release drug delivery device for diabetic treatment	Glipizide	HME, PVA	Producing a Duo Tablet with the controlled-release profile in a single process	[59]
FDM 3D printing	Immediate-release tablets containing thermolabile drug	Pantoprazole sodium	PVP	Reducing the infill rate while printing increased the tablets' porosity, which increased the dissolving rate.	[60]
FDM 3D printing at a lower printing temperature	Immediate release	Ramipril	Kollidon VA64 and Kollidon 12PF	Produce thermolabile and low-melting-point drugs when the drug is mixed with appropriate excipients	[61]
FDM 3D printing	Rapid release tablet	Haloperidol	Kollidon VA64, Kollicoat IR, Affinsiol TM 15 cP, and HPMCAS	Used to fabricate immediate-release dosage forms	[62]
FDM 3D printing	Drug-loaded implants	Quinine	Eudragit RS, Polycaprolactone (PCL), poly(L-lactide) (PLLA), and ethyl cellulose (EC)	Create a product with nearly any shape that can be made to work as an implant and contain the desired drug delivery mechanism at the same time.	[63]
FDM 3D printing	Drug-loaded implants	Indomethacin	Ethylene-vinyl acetate (EVA)	To create T-shaped intrauterine systems (IUS) and subcutaneous rods (SR) on demand.	[64]

TABLE 1.1 *(Continued)*

Manufacturing Method	Dosage Form	API	Excipients	Effects	References
FDM 3D printing	Drug-loaded implants	Indomethacin	PCL	Produce drug-containing T-shaped prototypes of IUS.	[65]
FDM 3D printing	Personalized drug-loaded vaginal rings.	Progesterone	Polyethylene Glycol (PEG) 4000	An effective method to prepare personalized and customized medications	[66]
FDM 3D printing	Bilayer dosage form	Metformin and Glimepride	Eudragit RL, polyvinyl alcohol (PVA)	It is possible to use two APIs simultaneously using the 3D-printed bilayer dosage form, which can be given to patients at different times of the day based on their needs.	[67]
HME coupled with FDM 3D printing	Chewable tablets	Indomethacin (IND)	HMPCAS.	Providing excellent taste masking efficiency by producing tablets with complex designs that might grab kids' interest.	[40]
FDM 3D printing	Orodispersible films (ODFs)	Aripiprazole	PVA	To accelerate the rate of dissolution of the weakly water-soluble substance.	[68]

1.5 CHALLENGES

One technique that shows promise for fast material manufacturing and prototyping is 3D printing. The new materials, like the original component, originate on a three-dimensional layer, much like in photocopying or printing. The field of pharmaceuticals has seen significant growth and advancement since the 1980s, but its use is still restricted [69]. A promising technology that can completely alter the pharmaceutical business is AM by enabling personalized medicine. However, several challenges need to be addressed before this technique is widely accepted. These challenges include standardizing printers, ensuring the availability of safe and usable materials, developing nondestructive analytical methods, dealing with regulatory issues, and mitigating potential unintended consequences [70]. The following points have been discussed in detail:

1. **Standardization of printers:**

   Ensuring that all 3D printers used in pharmaceutical applications comply with uniform standards is crucial for achieving consistent product quality and safety.

2. **Availability of safe and usable materials:**

   The components that are used in 3D printing for drugs must not only be safe for human consumption but also be capable of effectively delivering medications.

3. **Development of nondestructive analytical methods:**

   Innovating ways to analyze 3D-printed pharmaceuticals without destroying them is essential for quality control and assurance.

4. **Navigating Regulatory Issues:**

   Adhering to stringent regulatory requirements is vital. Any pharmaceutical product must meet health and safety standards before it can be marketed.

1.6 TECHNICAL ASPECTS AND CHALLENGES IN 3D PRINTING

1.6.1 ERRORS AND DEFECTS IN UNPROCESSED FEED INGREDIENTS

As seen in Figure 1.5, the method of gear feeding comprises two cylinders that have a tooth inlay. Usually, a screw spring that adjusts the tension on the extruding gears is used to hold them in place. Because each material has a distinct hardness and density, this tension enables the gears to increase or decrease friction on the filament, allowing for ease of movement when utilizing different materials (PEEK, ABS, PLA, Carbon Fibre, and so forth) [71].

adjusts the tension on the extruding gears in used to hold feeding ingredients in place.

FIGURE 1.5 Gear feeding procedure by screw spring.

Source: Adapted from Ref. [73], © Copyright 2022 by the authors. Licensee MDPI, Basel, Switzerland. https://creativecommons.org/licenses/by/4.0/

Two distinct feeding techniques are used in 3D printing. The most popular is Bowden extrusion, in which the printer's body houses the gears that feed the filament, and a tube transports the filament to the thermal coupling hot end. Figure 1.6 shows an illustration of a 3D filament printer that makes use of Bowden extrusion. This lessens the chance of thermocouple burnout and fracture.

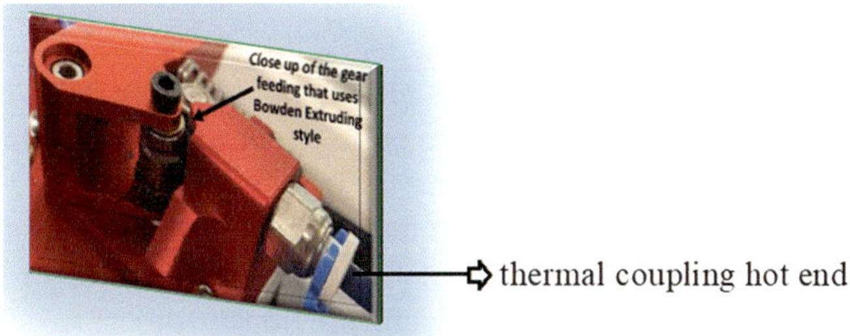

FIGURE 1.6 Close-up of the Bowden extrusion-style gear feeding apparatus.

Source: Adapted from Ref. [73], © Copyright 2022 by the authors. Licensee MDPI, Basel, Switzerland. https://creativecommons.org/licenses/by/4.0/

The thermocouple hot ends are in front of the extrusion gears of direct extrusion printers, or DXP printers. Because there is less tension on the filament and less time spent traveling between the gears, the thermocouple item has a reduced rate of failure. But because of the slop the gear motors produce, the print quality is decreased [72]. Both extruding systems frequently have an eye or a light sensor attached behind the extruding gears to notify the printer when the filament runs out. This then pauses the printing process and allows it to cool [73].

1.6.2 MACHINE ERRORS AND DEFECTS (3D PRINTERS)

1.6.2.1 BLOCKAGE OF NOZZLES

Numerous materials could clog printer nozzles during printing. The filament diameter, foreign matter, and substance mixing are some of the most frequent reasons why nozzles malfunction.

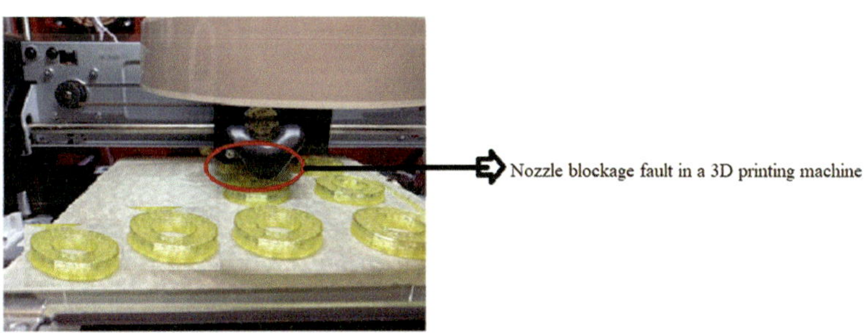

FIGURE 1.7 An issue about nozzle blockages in a 3D printer.

Source: Adapted from Ref. [73], © Copyright 2022 by the authors. Licensee MDPI, Basel, Switzerland. https://creativecommons.org/licenses/by/4.0/

1.6.2.2 BLOCKAGE OF FOREIGN MATTER

There are numerous potential causes of material blocking, so it is important to pay less attention to the printer itself and more attention to the environment it lives in. Tape is among the most widely used. Tape is used by several filament makers to attach the filament to the roll and start the coiling process. This leads to one of two problems because of the light sensors for the gear feeding that were previously described. The first problem is that the tape does not come off from the roll or the filament. This unintentionally results in poor print, the gear biting on the filament, and stress on the machine. The second problem is that although the tape comes off the roll, it sticks to the filament roll [74].

1.6.3 MISTAKES AND DEFECTS OCCURRING DURING PRINTING

When a feeding tube becomes clogged with tape, the obstruction requires replacing the entire tube, which must be removed. It is required to do a "hot push" of material if the tape has obstructed the extrusion nozzle. This involves pushing the filament through by hand after heating the extruder's thermocouple to the upper end of the filament's temperature range. This causes the tape and filament to melt down and be forcibly taken out of the system. After that, the foreign material should be removed from the extruder, theoretically Failure and Flaws in Printed Parts [75]**.** Errors and shortcomings may be exceedingly expensive and lengthy because of the nature of AM. The majority can be remelted and then replicated in material removal operations. Because of the nature of the processes involved in printing, any failure that occurs either inside the part or during the process would render it unusable [76].

1.6.3.1 SPAGHETTING

Failure known as "spaghetting" happens when the material being extruded sticks to the extrusion nozzles rather than connecting or adhering to the build plate (Figure 1.8). This failure is known by an unofficial moniker, although it is a regular problem that wastes a lot of filaments. The presence of foreign matter on the extruder tip is one of the reasons for this failure. Usually, this is the result of old filament from previous printing burning onto the nozzle. Dragging is an additional option. This occurs when the extrusion nozzle is not adjusted to account for potential variations in print bed height. After that, some of the build plate material is removed by dragging the nozzle across the build plate. Following that, this melts and sticks to the extrusion nozzle [77].

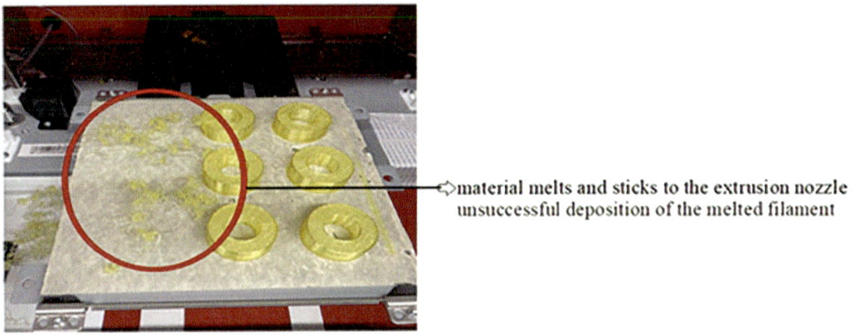

FIGURE 1.8 Spaghetting of the melted filament material's ineffective deposition.

Source: Adapted from Ref. [73], © Copyright 2022 by the authors. Licensee MDPI, Basel, Switzerland. https://creativecommons.org/licenses/by/4.0/

1.6.3.2 ADHESION AND WARPING

Print bed adhesion is often the most prevalent adhesion failure. The initial layers of the print, known as the print bed adhesion, help to maintain the print's stable position. The print will not proceed if the bed adheres. It is advised to utilize a hot and mildly abrasive surface to lower the likelihood of bed adhesion failure. As a result, the filament will experience friction, increasing the surface area contact and the likelihood of adhesion. The material that has been extruded is softened by heated beds, facilitating the adherence of the top layer with a rough surface to the initial layer of material [78, 79]. Another problem is layer adhesion. This happens when the material that was extruded fails to melt and correctly cling to the layer that came before it [80].

1.6.3.3 DENSITY AND POROSITY REGULARITY

Measuring material properties involves two primary methods. Density is defined as the mass per unit of volume, whereas porosity is the measurement of the space inside a portion that is devoid of material.

Similar to casting, trapped air bubbles can form during the printing process in AM; there are times when a bubble is injected into the material during the print process because the material is being extruded. There are little spaces between the layers and walls since prints are made linearly. The prints have porosity as a result of these micro gaps. Porosity changes consequently have an impact on density as well. The porosity and density of each cross-section of the printed component change if it is divided into separate portions. Fatigue is a result of voyeurism, an issue that is not widely explored. The relationship between voids and fatigue behavior is examined in this 2022 article [81]. The impulse excitation technique to find Young's modulus is a way to remedy this problem [82]. This procedure allows one to quantify the sample's inherent frequencies and subsequently compute its mechanical properties. It was discovered that the porosity problems might be resolved by optimizing the print's geometry after calculating Young's modulus. In composites, print optimization is also achievable. From the literature it was found that changing the printing parameters on the interfacial bond strength of ABS/carbon fiber-reinforced poly-lactic acid (CF-PLA) also changed the quality of the print [83].

1.6.3.4 SURFACE TOPOGRAPHY DIFFERENCES

As a result, the surface topography varies among components of the layer printing process employed by FDM. The printer extrudes partially melted material onto the substrate or on top of layers that have already been deposited. As previously mentioned in the chapter, this can lead to the material breaking during printing or air bubbles getting stuck between layers. Variations in topography may result in the final component if this occurs. Although the majority of the inaccuracies are within tolerances, occasionally there are significant variations in the topography. This may be the result of air pockets being introduced on the surface of the print degrading and collapsing on itself [84]. Further research into materials and production processes is required to improve our comprehension of the flaws in 3D-printed goods as well as possible malfunctions in 3D-printing equipment. Some of the suggested future work projects include specific experimental investigations into the causes of defects in 3D-printed polymers and composites; studies on process parameters and their influence on faults and failures; studies on machine failure cases and the implementation of potential feedback controls to prevent such failures; and studies on manufacturing process optimization toward the lowest possibility of defect generation [85, 86].

1.7 CONCLUSION

AM technology has shown incredible growth in different applications. One of the most popular AM methods is FDM, which is economical and can create parts with complex shapes. After conducting a thorough analysis, this chapter has concluded that various parameters in the FDM process, such as the specific machine and material parameters used, significantly affect the quality and mechanical properties of the final printed parts. It is crucial to consider these factors when aiming to optimize the FDM process to produce components that are of high quality and reliability.

Ongoing advancements and rigorous quality control measures in 3D printing have the potential to greatly impact the development, production, and administration of medication, leading to more personalized and effective healthcare solutions. In recent years, there has been a significant rise in the use of FDM technology to cater to the demand for customized medications, patient-centric therapies, personalized drug delivery, precision medicine, pediatrics, geriatrics, patient-specific implants and prosthetics, bio-robotics, wound dressing, dentistry, novel corona-virus drug development research, tissue engineering, drug testing, pharmacokinetic modeling, and simulation, as well as taste masking. However, several challenges must be addressed to facilitate the widespread adoption of this technology. Standardizing printers, ensuring the availability of safe materials, developing nondestructive analytical methods, navigating regulatory hurdles, and addressing potential unintended consequences are key challenges that demand attention. However, to fully realize its benefits, we need to address the challenges it poses and conduct further research to better understand the relationships between different factors and parameters. This will ensure a successful integration of AM into various industries.

KEYWORDS

- **3D printing method**
- **additive manufacturing**
- **fused deposition modeling**
- **personalized medicine**
- **DoE**
- **FDM**

REFERENCES

1. Andrews, G. P., Jones, D. S., Diak, O. A., McCoy, C. P., Watts, A. B., McGinity, J. W. (2008). The manufacture and characterization of hot-melt extruded enteric tablets. *European Journal of Pharmaceutics and Biopharmaceutics, 69*, 264–273.
2. Mwema, F. M., & Akinlabi, E. T. (2020). Basics of fused deposition modelling (FDM). In *Springer briefs in applied sciences and technology* (pp. 1–15). https://doi.org/10.1007/978-3-030-48259-6_1
3. Vyavahare, S., Teraiya, S., Panghal, D., & Kumar, S. (2020). Fused deposition modeling: A review. *Rapid Prototyping Journal*, *26*(1), 176–201. https://doi.org/10.1108/RPJ-04-2019-0106.
4. Varghese, R., Salvi, S., Sood, P., Karsiya, J., & Kumar, D. (2022). Recent advancements in additive manufacturing techniques employed in the pharmaceutical industry: A bird's eye view. *Annals of 3D Printed Medicine*, *8*, 100081. https://doi.org/10.1016/j.stlm.2022.100081
5. Tan, D. K., Maniruzzaman, M., & Nokhodchi, A. (2018). Advanced pharmaceutical applications of hot-melt extrusion coupled with fused deposition modeling (FDM) 3D printing for personalized drug delivery. *Pharmaceutics*, *10*(4). https://doi.org/10.3390/pharmaceutics10040203
6. Lalegani Dezaki, M., Mohd Ariffin, M. K. A., & Hatami, S. (2021). An overview of fused deposition modeling (FDM): Research, development and process optimization. In *Rapid Prototyping Journal, 27*(3), 562–582. https://doi.org/10.1108/RPJ-08-2019-0230
7. Bruce, L. D., Shah, N. H., Malick, A. W., Infeld, M. H., McGinity, J. W. (2005). Properties of hot-melt extruded tablet formulations for the colonic delivery of 5-aminosalicylic acid. *European Journal of Pharmaceutics and Biopharmaceutics, 59*, 85–97.
8. Dehghan-Manshadi, A., Yu, P., Dargusch, M., StJohn, D., & Qian, M. (2020). Metal injection molding of surgical tools, biomaterials, and medical devices: A review. *Powder Technology 364*, 189–204.

9. Maurya, R. K., & Niranjan, M. S. (2020). An experimental analysis of process parameters for EN-36C alloy steel using CNC lathe—A review. *Materials Today Proceedings, 25*, 773–777.
10. Doungkom, P., & Jiamjiroch, K. (2019). Analysis of printing pattern and infiltration percent over the tensile properties of PLA printed parts by a fuse deposition modelling printer. *IOP Conference Series Materials Science and Engineering, 501*, 012028.
11. Desu, K. P., Maddiboyina, B., Vanitha, K., Rao Gudhanti, N. K. S., Anusha, R., & Jhawat, V. (2021). 3D printing technology in pharmaceutical dosage forms: Advantages and challenges. *Current Drug Targets, 22*(16), 1901–1914. https://doi.org/http://dx.doi.org/10.2174/1389450122666210120142416
12. Manapat, J. Z., Chen, Q., Ye, P., & Advincula, R. C. (2017). 3D printing of polymer nanocomposites via stereolithography. *Macromolecular Materials and Engineering, 302*(9), 1–13. https://doi.org/10.1002/mame.201600553
13. Fina, F., Goyanes, A., Gaisford, S., & Basit, A. W. (2017). Selective laser sintering (SLS) 3D printing of medicines. *International Journal of Pharmaceutics*, *529*(1–2), 285–293. https://doi.org/10.1016/j.ijpharm.2017.06.082
14. Pilipovic, A., Sercer, M., Stojanovic, I., Brajlih, T., Drstvensek, I., & Valentan, B. (2011). Laminate object manufacturing vs. fused deposition modeling - Machine comparison. *Annals of DAAAM and Proceedings of the International DAAAM Symposium*, *22*(1), 221–222. https://doi.org/10.2507/22nd.daaam.proceedings.110
15. Miao, Z., Seo, J., & Hickner, M. A. (2018). Solvent-cast 3D printing of polysulfone and polyaniline composites. *Polymer*, *152*, 18–24. https://doi.org/10.1016/j.polymer.2018.05.055
16. Mwema, F. M., & Akinlabi, E. T. (2020). Basics of fused deposition modelling (FDM). In *Springer briefs in applied sciences and technology* (pp. 1–15). https://doi.org/10.1007/978-3-030-48259-6_1
17. Gavali, V. C., Kubade, P. R., & Kulkarni, H. B. (2020), Property enhancement of carbon fiber reinforced polymer composites prepared by fused deposition modeling. *Materials Today Proceedings, 23*, 221–229.
18. Dey, A., & Yodo, N. (2019). A systematic survey of FDM process parameter optimization and their influence on part characteristics. *Journal of Manufacturing and Materials Processing, 3*(3), 64.
19. Shergill, K., Chen, Y., & Bull, S. (2023). An investigation into the layer thickness effect on the mechanical properties of additively manufactured polymers: PLA and ABS. *International Journal of Advanced Manufacturing Technology*, *126*(7–8), 3651–3665. https://doi.org/10.1007/s00170-023-11270-y
20. Camargo, J.C., Machado, Á.R., Almeida, E.C., & Silva, E.F.M.S. (2019). Mechanical properties of PLAgraphene filament for FDM 3D printing. *The International Journal of Advanced Manufacturing Technology, 103*(5–8), 2423–2443.
21. Sharma, M., Sharma, V., & Kala, P. (2019). Optimization of process variables to improve the mechanical properties of FDM structures. *Journal of Physics: Conference Series*, *1240*(1). https://doi.org/10.1088/1742-6596/1240/1/012061
22. Maqsood, N., & Rimašauskas, M. (2022). Influence of printing process parameters and controlled cooling effect on the quality and mechanical properties of additively manufactured CCFRPC. *Composites Communications*, *35*, 101338. https://doi.org/https://doi.org/10.1016/j.coco.2022.101338.

23. Qamar Tanveer, M., Mishra, G., Mishra, S., & Sharma, R. (2022). Effect of infill pattern and infill density on the mechanical behavior of FDM 3D printed Parts- a current review. *Materials Today: Proceedings*, *62*, 100–108. https://doi.org/https://doi.org/10.1016/j.matpr.2022.02.310.
24. Giri, J., Patil, A., & Prabhu, H. (2019). The effect of various parameters on the nozzle diameter and 3D printed product in fused deposition modelling: An approach. In C. R. Krishna, M. Dutta & R. Kumar (Eds.), *Proceedings of 2nd International Conference on Communication, Computing and Networking* (pp. 839–847). Springer Singapore.
25. Alzyod, H., & Ficzere, P. (2023). Material-dependent effect of common printing parameters on residual stress and warpage deformation in 3D printing: A comprehensive finite element analysis study. *Polymers*, *15*(13), 2893. https://doi.org/10.3390/polym15132893.
26. Haque, S. S. (2020). *Minimizing Stringing Issues in FDM Printing*. https://www.researchgate.net/publication/346563712
27. Zhai, C., Wang, J., (Paul) Tu, Y., Chang, G., Ren, X., & Ding, C. (2023). Robust optimization of 3D printing process parameters considering process stability and production efficiency. *Additive Manufacturing*, *71*, 103588. https://doi.org/https://doi.org/10.1016/j.addma.2023.103588.
28. Long, J., Gholizadeh, H., Lu. J., Bunt, C., & Seyfoddin, A. (2017). Application of fused deposition modelling (FDM) method of 3D printing in drug delivery. *Current Pharmaceutical Design, 23*(3), 433–439.
29. Tan, D. K., Maniruzzaman, M., & Nokhodchi, A. (2018). Advanced pharmaceutical applications of hot-melt extrusion coupled with fused deposition modelling (FDM) 3D printing for personalised drug delivery. *Pharmaceutics, 10*, 203.
30. Singh, S., & Singh, R. (2017). Integration of fused deposition modeling and vapor smoothing for biomedical applications. In *Reference module in materials science and materials engineering* (pp. 1–15). Elsevier,
31. Goyanes, A., Scarpa, M., Kamlow, M., & Gaisford, S. (2017). Patient acceptability of 3D printed medicines. *International Journal of Pharmaceutics, 530*(1–2), 71–78.
32. Jamroz, W., Szafraniec, J., Kurek, M., et al. (2018). 3D printing in pharmaceutical and medical applications – recent achievements and challenges. *Pharmaceutics Research, 35*, 176
33. Alomari, M., Mohamed, F. H., Basit, A. W., et al. (2015). Personalized dosing: Printing a dose of one's own medicine. *International Journal of Pharmaceutics, 494*, 568–577.
34. The Lancet. (2018). Personalized medicine in the UK. *Lancet, 391*(10115), e1. doi: 10.1016/S0140-6736(17)33261-0. Epub 2017 Dec 21. PMID: 29275870.
35. Mwema, F. M., & Akinlabi, E. T. (2020). *Fused deposition modeling: Strategies for quality enhancement*. Springer Nature.
36. Januskaite, P., Xu, X., Ranmal, S. R., et al. (2020). I spy with my little eye: A paediatric visual preferences survey of 3D printed tablets. *Pharmaceutics, 12*, 1100.
37. Tabriz, A. G., Fullbrook, D. H. G., Vilain, L., Derrar, Y., Nandi, U., Grau, C., Morales, A., Hooper, G., Hiezl, Z., & Douroumis, D. (2021). Personalised tasted masked chewable 3D printed fruit-chews for paediatric patients. *Pharmaceutics, 13*(8), 1301
38. Quodbach, J., Bogdahn, M., Breitkreutz, J., et al. (2022). Quality of FDM 3D printed medicines for pediatrics: Considerations for formulation development, filament

extrusion, printing process and printer design. *Therapeutic Innovation & Regulatory Science, 56*, 910–928.
39. Karavasili, C., Gkaragkounis, A., Moschakis, T., et al. Pediatric-friendly chocolate-based dosage forms for the oral administration of both hydrophilic and lipophilic drugs fabricated with extrusion-based 3D printing. *European Journal of Pharmaceutical Sciences,147*, 105291
40. Scoutaris, N., Ross, S. A., & Douroumis, D. (2018). 3D printed 'Starmix' drug loaded dosage forms for paediatric applications. *Pharmaceutical Research, 35*(2), 16.
41. Maher, R. L., Hanlon, J., & Hajjar, E. R. (2013). Clinical consequences of polypharmacy in the elderly. *Expert Opinion on Drug Safety, 13*, 57–65.
42. Wang, S., Chen, X., Han, X., Hong, X., Li, X., Zhang, H., Li, M., Wang, Z., & Zheng, A. (2023). A review of 3D printing technology in pharmaceutics: Technology and applications, now and future. *Pharmaceutics, 15*, 416.
43. Basit, A. W., & Gaisford, S. (2018). *3D printing of pharmaceuticals*. Springer International Publishing.
44. Zuniga, J., Katsavelis, D., Peck, J., Stollberg, J., Petrykowski, M., Carson, A., et al. Cyborg beast: A low-cost 3D-printed prosthetic hand for children with upper-limb differences. *BMC Research Notes, 8*, 10.
45. Feinberg, A. W. (2015). Biological soft robotics. *Annual Review of Biomedical Engineering, 17*, 243–265.
46. Nawroth, J. C., Lee, H., Feinberg, A. W., Ripplinger, C. M., McCain, M. L., Grosberg, A., et al. (2012). A tissue-engineered jellyfish with biomimetic propulsion. *Nature Biotechnology, 30*, 792–727
47. Bracaglia, L. G., Messina, M. J., Winston. S., Kuo, C.-Y., Lerman, M., & Fisher, J. P. (2017). Printed pericardium hydrogels to promote wound healing in vascular applications. *Biomacromolecules, 18*(11), 3802–3811.
48. Khorsandi, D., Fahimipour, A., Abasian, P., Saber, S. S., et al. (2021). 3D and 4D printing in dentistry and maxillofacial surgery: Printing techniques, materials, and applications. *Acta Biomaterials, 122*, 26–49.
49. Muta, S., Ikeda, M., Nikaido, T., Sayed, M., Sadr, A., Suzuki, T., & Tagami, J. (2020). Chairside fabrication of provisional crowns on FDM 3D-printed PVA model. *Journal of Prosthodontic Research, 64*, 401–407.
50. Oladapo, B. I., Ismail, S. O., Afolalu, T. D., Olawade, D. B., & Zahedi, M. (2021). Review on 3D printing: Fight against COVID-19. *Materials Chemistry and Physics, 258*, 123943.
51. Cader, H. K., Rance, G. A., Alexander, M. R., Gonçalves, A. D., Roberts, C. J., Tuck, C. J., & Wildman, R. D. (2019). Water-based 3D inkjet printing of an oral pharmaceutical dosage form. *International Journal of Pharmaceutics, 564*, 359–368.
52. Vanderburgh, J., Sterling, J. A., & Guelcher, S. A. (2017). 3D printing of tissue-engineered constructs for in vitro modeling of disease progression and drug screening. *Annals of Biomedical Engineering, 45*, 164–179.
53. Olson, H., Betton, G., Robinson, D., Thomas, K., Monro, A., Kolaja, G., Lilly, P., Sanders, J., Sipes, G., Bracken, W., Dorato, M., Van Deun, K., Smith, P., Berger, B., & Heller, A. (2000). Concordance of the toxicity of pharmaceuticals in humans and animals. *Regulatory Toxicology and Pharmacology, 32*, 56–67.

54. Cailleaux, S., Sanchez-Ballester, N. M., Gueche, Y. A., Bataille, B., & Soulairol, I. (2021). Fused deposition modeling (FDM), the new asset to produce tailored medicines. *Journal of Controlled Release, 330*, 821–841.
55. Tabriz, A. G., Fullbrook, D. H. G., Vilain, L., Derrar, Y., Nandi, U., Grau, C., Morales, A., Hooper, G., Hiezl, Z., & Douroumis, D. (2021). Personalised tasted masked chewable 3D printed fruit-chews for paediatric patients. *Pharmaceutics, 13*, 1301.
56. Tabriz, A. G., Nandi, U., Scoutaris, N., Sanfo, K., Alexander, B., Gong, Y., Hui, H.-W., Kumar, S., & Douroumis, D. (2022). Personalised paediatric chewable ibuprofen tablets fabricated using 3D micro-extrusion printing technology. *International Journal of Pharmaceutics, 626*, 122135.
57. Chai, X., Chai, H., Wang, X., Yang, J., Li, J., Zhao, Y., & Xiang, X. (2017). Fused deposition modeling (FDM) 3D printed tablets for intragastric floating delivery of domperidone. *Scientific Reports, 7*(1), 2829.
58. Okwuosa, T. C., Pereira, B. C., Arafat, B., Cieszynska, M., Isreb, A., & Alhnan, M. A. (2017). Fabricating a shell-core delayed release tablet using dual FDM 3D printing for patient-centred therapy. *Pharmaceutical Research, 34*, 427–437.
59. Li, Q., Wen, H., Jia, D., Guan, X., Pan, H., Yang, Y., Yu, S., Zhu, Z., Xiang, R., & Pan, W. (2017). Preparation and investigation of controlled-release glipizide novel oral device with three-dimensional printing. *International Journal of Pharmaceutics, 525*, 5–11.
60. Kempin, W., Domsta, V., Grathoff, G., Brecht, I., Semmling, B., Tillmann, S., Weitschies, W., & Seidlitz, A. (2018). Immediate release 3D-printed tablets produced via fused deposition modeling of a thermo-sensitive drug. *Pharmaceutical Research, 35*, 124.
61. Kollamaram, G., Croker, D. M., Walker, G. M., Goyanes, A., Basit, A. W., & Gaisford, S. (2018). Low temperature fused deposition modeling (FDM) 3D printing of thermolabile drugs. *International Journal of Pharmaceutics, 545*, 144–152.
62. Solanki, N. G., Tahsin, M., Shah, A. V., & Serajuddin, A. T. M. (2018). Formulation of 3D printed tablet for rapid drug release by fused deposition modeling: Screening polymers for drug release, drug-polymer miscibility and printability. *Journal of Pharmaceutical Sciences, 107*, 390–401.
63. Kempin, W., Franz, C., Koster, L. C., Schneider, F., Bogdan, M., Weitschies, W., & Seidlitz, A. (2017). Assessment of different polymers and drug loads for fused deposition modeling of drug-loaded implants. *European Journal of Pharmaceutics and Biopharmaceutics, 115*, 84–93.
64. Genina, N., Holländer, J., Jukarainen, H., Mäkilä, E., Salonen, J., & Sandler, N. (2016). Ethylene vinyl acetate (EVA) as a new drug carrier for 3D printed medical drug delivery devices. *European Journal of Pharmaceutics and Sciences, 90*, 53–63.
65. Hollander, J., Genina, N., Jukarainen, H., Khajeheian, M., Rosling, A., Mäkilä, E., & Sandler, N. (2016). Three-dimensional printed PCL-based implantable prototypes of medical devices for controlled drug delivery. *Journal of Pharmaceutical Sciences, 105*, 2665–2676
66. Fu, J., Yu, X., & Jin, Y. (2018). 3D printing of vaginal rings with personalized shapes for controlled release of progesterone. *International Journal of Pharmaceutics, 539*, 75–82.

67. Gioumouxouzis, C. I., Baklavaridis, A., Katsamenis, O. L., Markopoulou, C. K., Bouropoulos, N., Tzetzis, D., & Fatouros, D. G. (2018). A 3D printed bilayer oral solid dosage form combining metformin for prolonged and glimepiride for immediate drug delivery. *European Journal of Pharmaceutical Sciences, 120*, 40–52.
68. Jamroz, W., Kurek, M., Łyszczarz, E., Szafraniec, J., Knapik-Kowalczuk, J., Syrek, K., Paluch, M., & Jachowicz, R. (2017). 3D printed orodispersible films with Aripiprazole. *International Journal of Pharmaceutics, 533*, 413–420.
69. Mateti, T., Jain, S., Ananda Shruthi, L., Laha, A., & Thakur, G. (2023). Chapter 1 - An overview of the advances in the 3D printing technology. In J. K. Pandey, S. Manna, R. K. Patel & M. Qian (Eds.), *3D printing technology for water treatment applications* (pp. 1–37). Elsevier. https://doi.org/https://doi.org/10.1016/B978-0-323-99861-1.00002-3
70. Desu, P. K., Maddiboyina, B., Vanitha, K., Gudhanti, S. N. K. R., Rapuri, A., & Jhawat, V. 3D printing technology in pharmaceutical dosage forms: Advantages and challenges. *Current Drug Targets, 22*, 1901–1914.
71. Ahmad, A. N., Gopinath, P., & Vinogradov, A. (2019). Chapter 1 - 3D Printing in medicine: Current challenges and potential applications. In N. Ahmad, P. Gopinath & R. Dutta (Eds.), *3D printing technology in nanomedicine* (pp. 1–22). Elsevier. https://doi.org/https://doi.org/10.1016/B978-0-12-815890-6.00001-3
72. Gibson, I., Rosen, D., Stucker, B., & Khorasani, M. (2020). *Additive manufacturing technologies*. Springer.
73. Baechle-Clayton, M., Loos, E., Taheri, M., & Taheri, H. (2022). Failures and flaws in fused deposition modeling (FDM) additively manufactured polymers and composites. *Journal of Composites Science*, *6*(7). https://doi.org/10.3390/jcs6070202
74. Groover, M. P. (2022). *Fundamentals of modern manufacturing*. Wiley
75. Georgantzinos, S. K. (2023). Characterization and modelling of composites, volume III. *Journal of Composites Science, 7*(11). https://doi.org/10.3390/jcs7110446
76. Grand View Research. (2018). Natural fiber composites (NFC) market size, share & trends analysis report by raw material, by matrix, by technology (injection molding, compression molding, pultrusion), by application, and by segment forecasts, 2018–2024; Report ID: 978-1-68038-890-9; Grand View Research: San Francisco, CA, USA.
77. Leary, M. (2020). Chapter 8 - Material extrusion. In M. Leary (Ed.), *Design for additive manufacturing* (pp. 223–268). Elsevier. https://doi.org/https://doi.org/10.1016/B978-0-12-816721-2.00008-7
78. Giles, H. F., Jr., Mount, E. M., III, Wagner, J. R., Jr. (2004). *Extrusion: The definitive processing guide and handbook*. William Andrew.
79. Griffey, J. (2017). Types of filaments for FDM printing. *Library Technology Reports, 53*, 12–16.
80. Bhusnure, O. G., Gholve, S. V., Sugave, B. K., Dongre, R. C., Gore, S. A., & Giram, P. S. (2016). 3D printing & pharmaceutical manufacturing: Opportunities and challenges. *International Journal of Bioassays, 5*(1), 4723–4738.
81. Badouard, C., Traon, F., Denoual, C., Mayer-Laigle, C., Paës, G., & Bourmaud, A. (2019). Exploring mechanical properties of fully compostable flax reinforced composite filaments for 3D printing applications. *Industrial Crops and Products, 135*, 246–250.

82. Gibson, I., Rosen, D. W., & Stucker, B. (2014). *Additive manufacturing technologies* (p. 17). Springer.
83. Leon-Becerra, J., González-Estrada, O. A., & Sánchez-Acevedo, H. (2022). Comparison of models to predict mechanical properties of FR-AM composites and a fractographical study. *Polymers (Basel)*, *14*(17), 3546. doi: 10.3390/polym14173546. PMID: 36080625; PMCID: PMC9460192.
84. Zhang, W., Zhang, Z., & Chen, X. (2012). Three-dimensional elastic wave numerical modeling in the presence of surface topography by a collocated-grid finite-difference method on curvilinear grids. *Geophysical Journal International*, *190*(1), 358–378. https://doi.org/10.1111/j.1365-246X.2012.05472.x
85. Ekberg, O., et al. (2002). Social and psychological burden of dysphagia; its impact on diagnosis and treatment. *Dysphagia, 17*, 139–146.
86. Ali, H., Kurokawa, S., Shehab, E., & Mukhtarkhan, M. (2023). Development of a large-scale multi-extrusion FDM printer, and its challenges. *International Journal of Lightweight Materials and Manufacture*, *6*(2), 198–213.

CHAPTER 2

Fundamental Principles of Fused Deposition Modeling in Pharmaceutical Context

SUPARNA GARAI, SAHANAZ KHATUN, MD. NAHID ZAMAN, PIJUSH KANTI SHIT, and AMALESH SAMANTA*

Division of Microbiology and Pharmaceutical Biotechnology, Department of Pharmaceutical Technology, Jadavpur University, Kolkata, West Bengal, India

*Corresponding author

ABSTRACT

Precision medicine has been gaining traction in conventional medicine during the past couple of years. Currently, conventional industrial methods produce patient-specific pharmaceutical formulations. Advanced 3D printing, specifically fused deposition modeling (FDM), offers a viable way to tailor the oral administration of medication. A form of extrusion-based technology called FDM 3D printing spreads the material across the build plate in successive layers as it emerges from the nozzle. Feeding material comes in the form of filaments, which are made by melting and extruding a physical combination of polymers and active pharmaceutical ingredients (API).

FDM is a technology used in the pharmaceutical sector to manufacture drug-loaded materials layer by layer. On the other hand, suitable material selection, nozzle diameter, types of filament selection, drug loading, layer thickness optimization, extrusion, bed temperature control, print resolution and accuracy, print bed and complement structures, print speed, infill

Advancements in Fused Deposition Modeling (FDM) Technology for Pharmaceutical and Medical Applications. Kuldeep Vinchurkar, Dinesh Kumar Mishra, Pankaj V Dixit, and Sheetal Mane (Eds.)

DOI: 10.1201/9781779640338-2

density and pattern, post-processing techniques, quality control and validation, and regulatory compliance are probably some of the prerequisite principles that controlled drug delivery systems taken into account. The requirement for an API to be included in the filament is critical in light of the prospective uses of FDM by the pharmaceutical sector. Printing the dose form depends significantly on the filament's characteristics, including consistent size, elasticity, and stiffness. Numerous research teams have devised and assessed a wide array of conventional and innovative dosage forms, including standard immediate-release tablets and capsules, controlled- and modified-release formulations, specialized floating systems for enhanced gastric retention, pulsatile release mechanisms, transdermal and transmucosal delivery systems, and tailor-made implants and devices. This approach has several benefits, including low cost, simple accessibility, and simplicity; however, it also has drawbacks, including unsatisfactory drug loading, a slow rate of drug dissolution and excipient thermal degradation.

2.1 INTRODUCTION

Fused deposition modeling (FDM) stands as a prevalent additive manufacturing (AM) technique, constructing three-dimensional objects through layer-by-layer material deposition. Regarded for its accessibility and versatility, FDM serves as a widely applicable 3D printing method, finding utility from rapid prototyping to producing final parts in diverse industrial sectors. FDM technology was developed in the late 1980s by Scott Crump, co-founder of Stratasys Inc., and has since become one of the most popular methods for 3D printing. In FDM, a thermoplastic filament is heated to its melting point and extruded through a nozzle onto a build platform. The nozzle moves along predetermined paths, depositing material layer-by-layer according to a computer-generated 3D model. As each layer cools and solidifies, the object gradually takes shape [1]. The FDM process involves several key components:

1. Extrusion nozzle: This is where the thermoplastic filament melted by heat is deposited on the build platform. Nozzles come in various sizes to accommodate different layer resolutions and material types.

2. Build platform: The build platform provides a stable surface upon which the object is built layer-by-layer. It may be heated to promote adhesion and prevent warping.
3. Filament: FDM printers use thermoplastic filaments, such as acrylonitrile butadiene styrene (ABS), polylactic acid (PLA), or polyethylene terephthalate glycol (PETG). These filaments come in different colors and properties, allowing for versatility in material selection.
4. Control system: A computerized control system interprets the 3D model data and directs the movement of the extrusion nozzle and build platform. This system ensures precise deposition of material according to the design specifications.
5. Prototyping: FDM is commonly used for rapid prototyping due to its ability to quickly and cost-effectively produce physical prototypes from digital designs. This enables engineers and designers to iterate and test their designs before moving to production.
6. Custom manufacturing: FDM is also used for custom manufacturing, allowing the building of unique or lower volume parts with complex geometries. This is particularly useful in industries such as aerospace, automotive, and healthcare.
7. Education and research: FDM is widely used in educational settings to teach students about 3D printing technology and design principles. It also plays a significant role in research, enabling scientists and engineers to explore new materials and applications in AM [2].

FDM technology offers a versatile and reasonable approach to 3D printing, allowing for the creation of complex objects with ease. With its widespread adoption across industries and applications, FDM continues to drive innovation and reshape the manufacturing landscape.

2.2 OVERVIEW OF PHARMACEUTICAL APPLICATIONS OF FDM

The three-dimensional printing (3DP) approach was initially discovered in the early 1980s and has since shown to be one of the incredibly inventive technologies, particularly in the pharmaceutical field. Over the past 10 years, improvements in pharmaceutical technologies have allowed for the personalization of treatments. The ability to produce intricate and personalized dose forms based on the demands of the patient has made 3DP a rapidly developing method in the production of pharmaceutical dosage forms.

2.2.1 DECISION

FDM provides unparalleled flexibility and precision in the production of pharmaceutical products, enabling us to tailor dosage forms to specific patient needs while optimizing manufacturing processes. Here's how FDM contributes to our decision-making processes:

1. **Rapid prototyping:** FDM facilitates rapid prototyping of dosage forms, allowing us to quickly iterate through design concepts and evaluate their feasibility. This accelerates the development cycle and enables us to make informed decisions based on tangible prototypes.
2. **Customization and personalization:** With FDM, we can customize dosage forms according to patient requirements, considering factors such as dosage strength, release kinetics, and patient preferences. This level of customization empowers us to make informed decisions regarding formulation design and patient-specific therapy.
3. **Complex geometries and multi-layered structures:** FDM excels in fabricating complex geometries and multi-layered structures with precision. This capability enables us to explore innovative dosage forms and drug delivery systems, leading to informed decisions about product design and performance.
4. **Cost-efficiency and resource optimization:** By minimizing material wastage and streamlining production processes, FDM contributes to cost-efficiency and resource optimization. This allows us to make informed decisions regarding manufacturing strategies and resource allocation, maximizing productivity while minimizing costs.
5. **Sustainability and environmental impact:** FDM offers environmental benefits by reducing energy consumption and material usage compared to traditional manufacturing methods.

Considering the growing importance of sustainability, FDM enables us to make informed decisions that align with our environmental goals. FDM empowers us to make informed decisions throughout the pharmaceutical manufacturing process, from formulation development to production. By leveraging the capabilities of FDM, we can optimize dosage form design, enhance patient outcomes, and drive innovation in the pharmaceutical field [3, 4].

2.2.2 DESIGN

2.2.2.1 TABLET

One popular form of 3D printing technology in many areas, including medicine, is fused deposition modeling, or FDM. FDM can be applied to many tablet manufacturing and development processes in the pharmaceutical industry. The following are a few medicinal uses for FDM in tablet manufacturing:

1. Prototype development: We can quickly prototype several tablet designs with FDM. Pharmacies can use conventional manufacturing processes after developing physical tablet forms based on their formulas.
2. Customized dosage forms: Dosage forms that are unique or customized can be made with FDM. This can be especially helpful when patients need certain dosages that are hard to find on the market.
3. Taste-masking and modified release: Drug release profiles can be altered or tastes can be concealed with the help of complicated geometries that can be fabricated with FDM. FDM can affect medication release kinetics and dissolving rates by modifying the internal structure of the tablet.
4. Combination dosage forms: FDM can create multi-layered tablets with chambers, making it possible to combine several medications or formulations into a single dosage form. For sequential drug release profiles or fixed-dose combo regimens, this may be beneficial.

5. Drug delivery systems: Tablets that adhere to the mucosa or dosage forms that are designed to be ingested by the stomach can be made using FDM.
6. Dose titration studies: The manufacturing of tablets with accurate and gradual adjustments in medication concentration is made easier by FDM. This capacity is useful for modifying dosage regimens to suit the needs of individual patients or for dose titration studies during medication development.
7. Paediatric and geriatric dosage forms: Tablets that meet the demands of both paediatric and geriatric populations can be made with precise sizes, shapes, and porosities, thanks to FDM. Tablets that are personalized can improve patient acceptance and simplicity of use.
8. Formulation screening: FDM is a tool that pharmaceutical researchers can use to make small batches of tablets for formulation optimization and screening investigations. This methodology expedites the process of developing new drugs by enabling quick iterations and assessment of various formulations [5, 6].

2.2.2.2 CAPSULE

1. Taste-masking and controlled release: Taste-masking chemicals or controlled-release formulations can be directly incorporated into the capsule shell, thanks to FDM technology. Pharmaceutical companies can create capsules with several compartments that contain distinct medication formulations for controlled release or combination therapy by manipulating the deposition of various materials.
2. Personalized medicine: Using FDM technology, customized medication formulations that meet the needs of each patient can be created. Pharmaceutical companies can create capsules with exact dosages and drug combinations that are optimized for individual patients by leveraging patient-specific data, such as genetic information or drug metabolism profiles.
3. Improved solubility and bio-availability: Additives to formulations that improve drug solubility and bio-availability can be

incorporated via FDM. Pharmaceutical companies can optimize drug delivery systems to improve therapeutic outcomes by carefully managing the deposition of these components.

4. Prototype development and clinical trials: Capsule formulations can be designed alliteratively and quickly with the use of FDM technology. Pharmaceutical firms can expedite the medication development process and save expenses connected with conventional manufacturing processes by using FDM to make small batches of capsules for preclinical testing and clinical trials.

2.2.3 DEVELOPMENT

2.2.3.1 CONTROLLED- AND DELAYED-RELEASE DOSAGE FORMS

On comparison with traditional administration, controlled-release doses prolong the duration of stable plasma drug concentrations and minimize the need for frequent doses. Both medication and patient compliance are improved by these manufactured tablets of ibuprofen that are 3D printed and have an interior scaffold structure for regulated medication release. The matrix polymer utilized was ethyl cellulose (EC) and the release modifiers that were assessed were polyvinyl alcohol (PVA), xanthenes gum, HPMC (K100-LV), and sodium alginate. The ibuprofen content in the formulation was adjusted to range from 16 to 24% (w/w) to assess how drug loading affected the *in vitro* release profiles and 3D printability. Tablets with varying infill densities (15, 20, and 25%) and the thicknesses of the shell (0.4, 0.8, and 1.2 mm) were printed to assess the impact of inner structure on rate of drug release. A 20% weight-to-weight medication load in the formulation yielded the best filaments for 3D printing. 95% of the drug was released from tablets containing 25% w/w HPMC in 24 h, as opposed to 17.8% in tablets without HPMC, according to drug release profiles. The cumulative drug release at 24 h rose with an increase in the formulation's drug concentration, while the quality of the printed tablet declined. The releasing performance of the tablets was significantly inversely proportional to their fill density and shell thickness. The authors concluded that a 24-h controlled-release formulation may be achieved by adjusting the formulation's composition as well as printing settings.

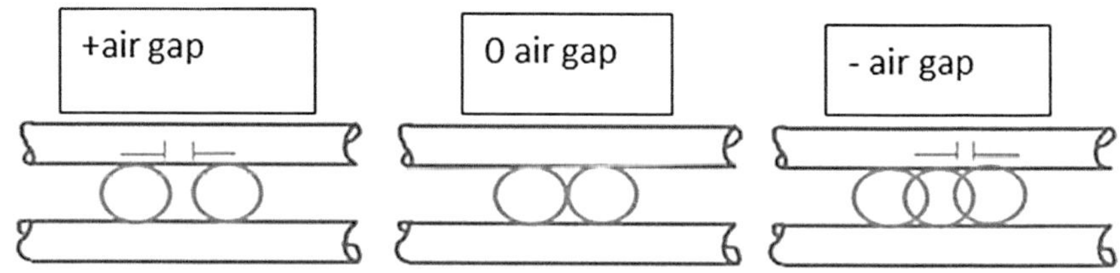

FIGURE 2.5 The air gap variable represented graphically.

9. **Nozzle diameter:** The road width is significantly impacted by the nozzle diameter. This is due to the fact that the nozzle diameter directly affects the decline in pressure through the liquefier [31]. In their experiments, Turner et al. [34] demonstrated that the nozzle's length-to-diameter (L/D) ratio influences the extent of pressure drop variation. It was noted that pressure drop increases as the nozzle diameter (D) decreases. Selecting the appropriate nozzle diameter is vital for maintaining a consistent and accurate flow of the extruded material. Additionally, it has been noted that the nozzle diameter significantly affects the extrusion time. For this reason, a thorough investigation must be performed to optimize the nozzle diameter value while accounting for all relevant process variables. A larger nozzle diameter correlates to a shorter extrusion time. Furthermore, it has been established that the measurement of the nozzle diameter significantly influences the geometric inaccuracy.
10. **Extruder temperature:** This represents the nozzle's temperature. This parameter has an impact on the polymer's viscosity. The study performed by Frank et al. [35] demonstrated that the shrinkage is significantly influenced by the high temperature of the nozzle. In 2015, investigation was performed on the influence of extruder temperature on the internal cavity and precision of FDM-generated geometries. According to the research findings as reported by Kaveh et al. [36], temperature significantly affects both output variables.
11. **Number of shells:** Every layer has an extruded boundary known as a shell. Each layer can have no less than one shell. The part can get stronger by adding more shells, without affecting its exterior size.

Griffiths et al. [37] took into account a variety of process characteristics, one of which being the number of shells.

12. **Retraction speed:** The rate at which the filament retracts is indicated by the retraction speed. This parameter is frequently overlooked in queries. However, as this study uses two pieces that are produced simultaneously, the retraction speed may have a substantial impact on the final integrity of the components and the duration of the printing process [20].
13. **Contour width and number of contours:** It is the breadth of the road route encircling the way's profile [38]. The flexural characteristics of the manufactured items are influenced by the contour numbers. The flexibility improves in direct proportion to a rise in contour number. This is a result of the raster count and related length being reduced, and it is observed that the load is delivered directly to the contour instead of the raster. Higher flexural characteristics were shown by the 900 raster angle and more outlines [39].

2.4.1 MATERIALS USED FOR FDM PROCESS

A significant challenge hindering the widespread adaption of the FDM technique in industry is the restricted availability of materials suitable for filament use. Currently, the most frequently utilized filament materials are ABS and PLA, each of which has unique properties [19]. While ABS has excellent ductility and is extremely heat resistant, PLA is biodegradable and has good stiffness. It is mostly unknown how the procedure limitations affect other materials. Rodríguez-Panes et al. [40] demonstrated that, in contrast to ABS, procedure parameter is important for PLA. Lately, studies have begun using natural fibers as additive materials into FDM filaments [41] and the level of fiber content varies from 20 to 30% [42].

2.4.2 STUDYING THE OPTIMIZATION OF FDM PROCESSES

The optimization of build time processes, surface roughness, dimensional accuracy, mechanical characteristics, and material behavior are all enhanced by research on FDM process settings. There has been discussion of crucial process variables that impact the processed part's quality. On this subject, a lot of research has been done, mostly concentrating on process optimization and experimental outcomes. In order to increase the surface finish, dimensional accuracy, and mechanical qualities of items processed with ABS, the majority of research on FDM process parameters has focused on process parameter optimization. Numerous scholars have proposed employing suitable statistical designs and optimization methods to investigate how process factors affect FDM processed products. Subsequent subsections provide a detailed review of the research on each quality parameters.

2.4.3 SURFACE ROUGHNESS

Anitha et al. [43] research focused on how several significant FDM process factors affected the ABS prototype's surface roughness. In this chapter, the signal to noise ratio (S/N), the Taguchi design matrix, and the analysis of variance (ANOVA) were employed. Road width, layer thickness, and deposition speed were the three process characteristics that were taken into account. Comparing layer thickness to road width and speed, this study found that layer thickness had the greatest impact on surface roughness. Additionally, it was shown that surface roughness and layer thickness had an inverse relationship.

Utilizing a statistical optimization approach that combines the Taguchi technique with gray relational analysis, Wang et al. [41] investigated the influence of control parameters on surface roughness. These parameters included layer thickness, deposition style, support style, and deposition orientations. Overall, a 62.27% enhancement in surface roughness was attained by employing the optimal factor settings. This study demonstrated that the Taguchi method, when compared to a complete factorial design, enabled the determination of optimal parameter combinations for surface roughness with fewer experiments, while still yielding comparable results.

2.4.4 DIMENSIONAL ACCURACY

Wang et al. [41] also noted that depositing thickness and construction orientation affected the produced part's dimensional correctness.

Different deposition patterns led to variations in dimensional precision in the various construction directions. According to Sood et al. [44], the gray Taguchi technique was used to investigate the effects of five process factors on the dimensional accuracy of an FDM-fabricated ABSP400 part. These parameters included part orientation, road width, layer thickness, air gap, and raster angle. They indicated that because the dimensions were either less or greater than the intended value, there was shrinkage along the length, width, and diameter of the hole in the manufactured part. By combining the Taguchi technique with fuzzy comprehensive assessment, Zhang and Peng [31] empirically established relationships between process parameters (extrusion velocity, filling velocity, layer thickness, and wire-width compensation) and dimensional error and deformation of FDM manufactured ABS part. According to their findings, layer thickness of 0.15 mm, extrusion velocity of 20 mm/s, filling velocity of 30 mm/s, and wire-width compensation of 0.17 mm were the ideal process parameter values for dimensional inaccuracy.

2.4.5 MATERIAL BEHAVIOR

To manufacture flexible ABS objects, Lee et al. [45] conducted an experimental investigation to optimize rapid prototyping parameters. They employed the ANOVA technique and the Taguchi method, considering factors such as layer thickness, raster angle, raster width, and air gap. In order to determine the part's elastic performance, the study found that three crucial elements were air gap, raster angle, and layer thickness. With a 0.18% error rate, the ideal parameters were identified and the outcomes were in good agreement with the lab tests. Using the Taguchi approach and ANOVA process, Abdullah et al. [46] evaluated the influence of air gap, raster angle, raster width, and layer thickness on the tensile properties of ABS material for FDM. Their analysis allowed them to determine the best possible combinations of process variables to increase overall performance in terms of elasticity. Zhang and Chou [47] examined the relationship between part deformities and stress distribution and process factors. The stress distribution and part deformities at various deposition circumstances were assessed using a finite element model that they built. To establish the correlation between process parameters and residual stresses as well as part distortions in the FDM process, central composite design (CCD) and ANOVA were utilized. Three key factors were chosen: road width, layer thickness, and scan speed. According to reports, the primary factor influencing residual stresses and part deformation was layer thickness. Following the inquiry, it was concluded that altering layer thickness and road width during the deposition phase increased residual stresses and part deformations. The primary finding of this study was that, with a slight inaccuracy, the finite element model that was built agreed well with the experimental findings. Nevertheless, one drawback of this research was that not all of the objectives could be concurrently satisfied by a single parameter adjustment.

2.4.6 BUILD TIME

Moreover, Thrimurthulu et al. [48] introduced a mathematical approach to predict and enhance the build time. Their analysis revealed that the most critical process variable affecting construction time was the build orientation. Using a real coded genetic algorithm (GA), they identified the optimal build orientation. Comparative analysis between the predictive capabilities of their model and previously published results demonstrated a reasonable agreement. In another study, Nancharaiah [49] used "Taguchi's design matrix L^9 orthogonal array" and the ANOVA technique to examine the association between process parameters and build time. It mentioned that the building time could be greatly impacted by process variables such as layer thickness and air gap. It also found that build time can be influenced by the layer thickness and air gap—66.57 and 30.77% of the total, respectively. The findings also showed that the ideal values for lowering the construction time were raster angle which measured 30, air gap of 0.020 mm, and layer thickness of 0.330 mm. Gurrula and Regalla [50] conducted a study using a full factorial design with `25 replications to investigate how different process parameters, such as layer thickness, raster angle, orientation, contour width, and part raster width, influence the volume of support material and the build time of FDM parts. Experiments revealed that crucial elements in minimizing the build time were the building direction and layer thickness. Still, the study did not concentrate on the best process parameters to reduce the volume of support material and build time.

2.5 REGULATORY CONSIDERATIONS IN PHARMACEUTICAL FDM

Over the past decade, personalized medicine has experienced significant growth in interest and comprehension. The United States Food and Drug Administration (FDA) granted approval for the first FDM printed drug product, marking a milestone in the pharmaceutical manufacturing sector [51]. However, stringent regulatory standards are essential to ensure that each product batch meets specifications concerning purity, impurities, and quality. The International Council for Harmonization (ICH) endeavors to achieve global consensus on the scientific and technical aspects of drug registration. The principles of quality, safety, efficacy, and multidisciplinary guidelines outlined by ICH can be applied to FDM technology for pharmaceutical products. These principles are governed by ICH Q8 (R2), Q9, Q10, and Q11 employing science and risk-based approaches. Other manufacturing and quality aspects of FDM drug products, including stability, impurities, drug substance and product specifications, and Good Manufacturing Practice (GMP) can adhere to ICH guidelines such as Q1, Q2, Q3, Q6, and Q7 [52].

Each region has established its own regulations and compendium pharmacopoeias to oversee pharmaceutical products. Should 3D printing be utilized, it should fall within the regulatory purview of each region for mass production and distribution within their respective countries.

2.6 QUALITY CONTROL AND ASSURANCE IN PHARMACEUTICAL FDM

Quality control and assurance in pharmaceutical FDM is important to assure the safety, efficacy, and consistency of 3D printed pharmaceutical products. FDM is a type of AM technology that involves depositing layers of material to build a 3D object based on a digital model. Within the pharmaceutical sector, FDM is experiencing growing utilization for crafting personalized dosage forms like tablets and implants. These forms offer precise dosing and customized release profiles. Here's a detailed overview of quality control and assurance in this context as described in the following sections.

2.6.1 MATERIAL CONTROL

The quality control process begins with the selection and testing of materials used in FDM. Pharmaceutical-grade polymers and excipients must meet stringent standards for purity, stability, and compatibility with the printing process. Raw material suppliers should provide certificates of analysis (CoA) verifying the material's properties and compliance with regulatory requirements.

2.6.2 PROCESS VALIDATION

Validation of the FDM process is essential to ensure that printed dosage forms consistently meet predetermined specifications. This involves establishing critical process parameters (CPPs) such as temperature, speed, layer thickness, and conducting comprehensive validation studies to demonstrate process robustness and reproducibility.

2.6.3 IN-PROCESS MONITORING

Real-time monitoring of key process parameters during printing helps detect deviations and maintain product quality. Sensors and monitoring systems can track parameters such as temperature, pressure, and filament diameter to ensure adherence to specifications. Any deviations should trigger immediate corrective actions to prevent the production of defective products.

2.6.4 QUALITY BY DESIGN (QBD)

Implementing a QbD approach involves systematically designing and controlling the FDM process to assure the quality of the product. This involves pinpointing critical quality attributes (CQAs) of the dosage form, conducting risk assessments, and fine-tuning process parameters to attain desired results while reducing variability.

2.6.5 MATERIAL TESTING

Before use in FDM, materials undergo rigorous testing to assess their physical, chemical, and mechanical properties. This includes characterization of polymer rheology, thermal behavior, degradation kinetics, and drug-polymer interactions. In addition, compatibility studies are conducted to evaluate the stability and performance of drug-loaded filaments during printing and storage.

2.6.6 POST-PRINTING INSPECTION

After printing, dosage forms undergo inspection to verify dimensional accuracy, surface finish, and structural integrity. Sophisticated imaging methods like microscopy, computed tomography (CT), and spectroscopy are employed to evaluate layer uniformity, porosity, and drug distribution within the dosage form.

2.6.7 QUALITY CONTROL TESTING

Finished dosage forms undergo comprehensive quality control testing to ensure compliance with pharmacopeial standards and regulatory requirements. This includes assays for drug content, dissolution testing to evaluate release kinetics, and physicochemical analyses to assess stability, hardness, and disintegration properties.

2.6.8 DOCUMENTED PROCEDURES AND RECORDS

All quality control and assurance activities are documented in standard operating procedures (SOPs) and batch records to ensure traceability and accountability. Records of material testing, process parameters, inspection results, and analytical testing are maintained for regulatory compliance and continuous improvement.

2.6.9 REGULATORY COMPLIANCE

Pharmaceutical FDM operations must adhere to regulatory guidelines and standards established by agencies such as the U.S. FDA and the European Medicines Agency (EMA). This includes compliance with GMP, submission of regulatory filings, and participation in inspections and audits.

2.6.10 CONTINUOUS IMPROVEMENT

Quality control and assurance processes should be periodically reviewed and optimized to incorporate new technologies, address emerging challenges, and improve overall efficiency and effectiveness. This may involve implementing advanced analytical techniques, automating quality control procedures, and adopting to industry best practices [53].

In conclusion, quality control and assurance in pharmaceutical FDM are essential to ensure the safety, efficacy, and consistency of 3D-printed dosage forms. By implementing robust quality management systems, adhering to regulatory requirements, and continuously improving processes, pharmaceutical manufacturers can confidently leverage FDM technology to produce high-quality personalized medications for patients.

2.7 CASE STUDIES AND EXAMPLES IN PHARMACEUTICAL FDM

Over the past decade, FDM technology has emerged as a cost-effective means for developing innovative drug delivery systems. Specifically, within polymeric drug delivery systems, FDM offers notable advantages. For instance, Scoutaris et al. [54] combined hot melt extrusion (HME) with FDM to fabricate drug-loaded filaments, resulting in tablets with reduced bitterness, immediate drug release within 60 min, and rapid dissolution rates. Similarly, Okwuosa et al. [55] employed 3D printing FDM and HME with a thermostable filler talc, to produce immediate release tablets at low temperatures. These tablets exhibited good mechanical properties, acceptable batch-to-batch variability, and immediate *in vitro* release patterns. This study notably marks the first reported use of polyvinylpyrrolidone (PVP) in formulating immediate release tablets.

Moreover, studies have explored the modification of dissolution profiles through 3D printing FDM methodology. Goyanes et al. [56] investigated the impact of utilizing 5-ASA and 4-ASA isomers on dissolution profiles, highlighting the influence of printing parameters and geometries. They found that adjusting the infill percentage led to improved dissolution profiles and faster drug release, particularly beneficial for fabricating low-dose formulations.

In a similar vein, Goyanes et al. [57] demonstrated the feasibility of manufacturing delayed-release matrix printlets targeting different gastrointestinal segments. By varying enteric-coated polymer grades and API loading, they achieved controlled drug release profiles dependent on polymer composition and printing parameters.

Furthermore, researchers have explored the fabrication of drug-loaded implants using FDM technology. Khaled et al. [58] successfully loaded paracetamol using FDP 3D printing, ensuring high-drug content and meeting international standards for physical and mechanical properties. Similarly, Gioumouxouzis et al. [59] developed a pH-responsive formulation for targeted delivery of cytostatic 5-fluorouracil in colorectal carcinoma, achieving efficient drug release and minimizing manufacturing defects.

Moreover, Stewart et al. [60] investigated the application of poly (caprolactone) (PCL)-based coatings on implanted devices to extend the release of hydrophilic drugs. By combining fused filament fabrication (FFF) AM with PCL and poly (ethylene glycol) (PEG) coatings, they achieved adaptable release rates suitable for various medications.

Overall, these studies highlight the versatility and potential of FDM technology in personalized drug delivery systems, offering tailored solutions for specific patient needs.

2.8 FUTURE DIRECTIONS AND CHALLENGES IN PHARMACEUTICAL FDM

1. Customized dosage forms: FDM allows for the creation of customized dosage forms designed to meet the specific requirements of individual patients. Future advancements may involve the integration of patient-specific data, such as medical history and genetic information, to design and manufacture dosage forms with optimized drug release profiles and dosages [57].
2. Complex geometries: Advances in FDM technology aim to overcome limitations in the fabrication of complex geometries for drug delivery systems. Integration of multi-material printing and support structures can facilitate the production of intricate dosage forms, including implants and transdermal patches, with enhanced functionality and performance.

3. Combination products: Incorporating various drugs or therapeutic agents into a single dosage form offers potential for improved therapeutic results and patient adherence. Future directions in pharmaceutical FDM may involve the development of multi-drug delivery systems, including combination tablets and co-encapsulated formulations, for the treatment of complex diseases [61].
4. Incorporation of functional excipients: Functional excipients play a crucial role in modulating drug release kinetics, stability, and bioavailability. Future advancements in pharmaceutical FDM may focus on the incorporation of functional excipients, such as polymers, surfactants, and nanoparticles into printed dosage forms to achieve desired pharmacokinetic profiles and therapeutic effects [62].
5. Regenerative medicine: FDM holds promise for the fabrication of scaffolds and tissue-engineered constructs for regenerative medicine applications. Future directions may involve the integration of bioactive materials and cell-laden hydrogels into printed scaffolds to promote tissue regeneration and organ repair.

2.8.1 CHALLENGES

2.8.1.1 MATERIAL SELECTION AND COMPATIBILITY

The availability of suitable materials with appropriate biocompatibility, mechanical properties, and printability remains a challenge in pharmaceutical FDM. Limited options for pharmaceutical-grade filaments and resins hinder the development of novel drug delivery systems and medical devices [57].

2.8.1.2 PRECISION AND RESOLUTION

Achieving precise control over printing parameters and resolution is essential for the fabrication of high-quality dosage forms with uniform drug distribution and release. Challenges related to layer adhesion, surface roughness, and printing accuracy may impact the performance and efficacy of printed pharmaceutical products [63].

2.8.1.3 REGULATORY CONSIDERATIONS

Regulatory requirements for Pharmaceutical FDM pose challenges in terms of material characterization, process validation, and product quality assurance. Ensuring compliance with regulatory standards set forth by agencies such as the FDA and EMA is critical for the approval and commercialization of printed pharmaceutical products.

2.8.1.4 SCALABILITY AND PRODUCTION COST

Scaling up pharmaceutical FDM processes to meet commercial demand while maintaining cost-effectiveness presents challenges in terms of equipment investment, production throughput, and material costs. Balancing scalability with product quality and affordability is essential for the widespread adoption of FDM in pharmaceutical manufacturing [13].

2.8.1.5 LONG-TERM STABILITY AND SAFETY

Ensuring the long-term stability and safety of printed pharmaceutical products is paramount to their clinical efficacy and patient acceptance. Challenges related to degradation, leaching of additives, and potential toxicological effects necessitate comprehensive stability testing and risk assessment throughout the product lifecycle [64].

2.9 CONCLUSION

In the current scenario in pharmaceutical field, FDM is in growing mode. It is expected that the problems which are faced in traditional manufacturing procedure, will be overcome with this modern technique.

KEYWORDS

- **drug loading**
- **dosage forms**
- **excipient**
- **prototype development**
- **plasticizer**
- **surface roughness**
- **thermoplastic filament**

REFERENCES

1. Crump, S. S., & Muir, A. E.-P. D. (1992). *Creating three-dimensional objects.*
2. Gibson, I., Rosen, D., & Stucker, B. (2015). *Additive manufacturing technologies: 3D printing, rapid prototyping, and direct digital manufacturing.* Springer. https://doi.org/10.1007/978-1-4939-2113-3
3. Kulkarni, V. R., Kazi, M., Shahba, A. A.-W., Radhanpuri, A., & Maniruzzaman, M. (2022). Three-dimensional printing of a container tablet: A new paradigm for multi-drug-containing bioactive self-nanoemulsifying drug-delivery systems (Bio-SNEDDSs). *Pharmaceutics*, *14*(5), 1082. https://doi.org/10.3390/pharmaceutics14051082
4. Melocchi, A., Parietti, F., Maroni, A., Foppoli, A., Gazzaniga, A., & Zema, L. (2016). Hot-melt extruded filaments based on pharmaceutical grade polymers for 3D printing by fused deposition modeling. *International Journal of Pharmaceutics*, *509*(1–2), 255–263. https://doi.org/10.1016/j.ijpharm.2016.05.036
5. Aho, J., Bøtker, J. P., Genina, N., Edinger, M., Arnfast, L., & Rantanen, J. (2019). Roadmap to 3D-printed oral pharmaceutical dosage forms: Feedstock filament properties and characterization for fused deposition modeling. *Journal of Pharmaceutical Sciences*, *108*(1), 26–35. https://doi.org/10.1016/j.xphs.2018.11.012
6. Dumpa, N., Butreddy, A., Wang, H., Komanduri, N., Bandari, S., & Repka, M. A. (2021). 3D printing in personalized drug delivery: An overview of hot-melt extrusion-based fused deposition modeling. *International Journal of Pharmaceutics*, *600*, 120501. https://doi.org/10.1016/j.ijpharm.2021.120501
7. Jamroz, ´ W., Kurek, M., Łyszczarz, E., Szafraniec, J., Knapik-Kowalczuk, J., Syrek, K., Paluch, M., & Jachowicz, R. (2017). 3D printed orodispersible films with Aripiprazole. *International Journal of Pharmaceutics, 533*(2), 413–420. https://doi.org/10.1016/j.ijpharm.2017.05.052.
8. Turk, D. C., O'Connor, A. B., Dworkin, R. H., Chaudhry, A., Katz, N. P., Adams, E. H., Brownstein, J. S., Comer, S. D., Dart, R., Dasgupta, N., Denisco, R. A., Klein,

M., Leiderman, D. B., Lubran, R., Rappaport, B. A., Zacny, J. P., Ahdieh, H., Burke, L. B., Cowan, P., Jacobs, P., Malamut, R., Markman, J., Michna, E., Palmer, P., PeirceSandner, S., Potter, J. S., Raja, S. N., Rauschkolb, C., Roland, C. L., Webster, L. R., Weiss, R. D., & Wolf, K. (2012). Research design considerations for clinical studies of abuse-deterrent opioid analgesics: IMMPACT recommendations. *Pain, 153*(10), 1997–2008. https://doi.org/10.1016/j.pain.2012.05.029.

9. Fu, J., Yu, X., & Jin, Y. (2018). 3D printing of vaginal rings with personalized shapes for controlled release of progesterone. *International Journal of Pharmaceutics, 539*(1-2), 75–82. https://doi.org/ 10.1016/j.ijpharm.2018.01.036
10. Charalabidis, A., Sfouni, M., Bergström, C., & Macheras, P. (2019). The biopharmaceutics classification system (BCS) and the biopharmaceutics drug disposition classification system (BDDCS): Beyond guidelines. *International Journal of Pharmaceutics*, *566*. https://doi.org/10.1016/j.ijpharm.2019.05.041
11. Mwema, F. M., & Akinlabi, E. T. (2020). Basics of fused deposition modelling (FDM). *Fused Deposition Modeling*, 1–15. https://doi.org/10.1007/978-3-030-48259-6_1
12. Kristiawan, R. B., Imaduddin, F., Ariawan, D., Ubaidillah, & Arifin, Z. (2021). A review on the fused deposition modeling (FDM) 3D printing: Filament processing, materials, and printing parameters. *Open Engineering*, *11*(1), 639–649. https://doi.org/10.1515/eng-2021-0063
13. Chai, X., Chai, H., Wang, X., Yang, J., Li, J., Zhao, Y., Cai, W., Tao, T., & Xiang, X. (2017). Fused deposition modeling (FDM) 3D printed tablets for intragastric floating delivery of domperidone. *Scientific Reports*, *7*(1), 2829. https://doi.org/10.1038/s41598-017-03097-x
14. Sadia, M., Sośnicka, A., Arafat, B., Isreb, A., Ahmed, W., Kelarakis, A., & Alhnan, M. A. (2016). Adaptation of pharmaceutical excipients to FDM 3D printing for the fabrication of patient-tailored immediate release tablets. *International Journal of Pharmaceutics*, *513*(1–2), 659–668. https://doi.org/10.1016/j.ijpharm.2016.09.050
15. Ilyés, K., Kovács, N. K., Balogh, A., Borbás, E., Farkas, B., Casian, T., Marosi, G., Tomuță, I., & Nagy, Z. K. (2019). The applicability of pharmaceutical polymeric blends for the fused deposition modelling (FDM) 3D technique: Material considerations–printability–process modulation, with consecutive effects on in vitro release, stability and degradation. *European Journal of Pharmaceutical Sciences*, *129*, 110–123. https://doi.org/10.1016/j.ejps.2018.12.019
16. Jaisingh Sheoran, A., & Kumar, H. (2020). Fused deposition modeling process parameters optimization and effect on mechanical properties and part quality: Review and reflection on present research. *Materials Today: Proceedings*, *21*, 1659–1672. https://doi.org/10.1016/j.matpr.2019.11.296
17. Ngo, T. D., Kashani, A., Imbalzano, G., Nguyen, K. T. Q., & Hui, D. (2018). Additive manufacturing (3D printing): A review of materials, methods, applications and challenges. *Composites Part B: Engineering*, *143*, 172–196. https://doi.org/10.1016/j.compositesb.2018.02.012
18. Stansbury, J. W., & Idacavage, M. J. (2016). 3D printing with polymers: Challenges among expanding options and opportunities. *Dental Materials: Official Publication of the Academy of Dental Materials*, *32*(1), 54–64. https://doi.org/10.1016/j.dental.2015.09.018

45. Lee, B. H., Abdullah, J., & Khan, Z. A. (2005). Optimization of rapid prototyping parameters for production of flexible ABS object. *Journal of Materials Processing Technology*, *169*(1), 54–61. https://doi.org/10.1016/j.jmatprotec.2005.02.259
46. Abdullah, J., Khan, Z., & S.Y, K. (2006). Optimizing flexible behaviour of bow prototype using taguchi approach. *Journal of Applied Sciences*, *6*. https://doi.org/10.3923/jas.2006.622.630
47. Zhang, Y., & Chou, K. (2008). A parametric study of part distortions in fused deposition modelling using three-dimensional finite element analysis. *Proceedings of The Institution of Mechanical Engineers Part B-Journal of Engineering Manufacture*, *222*. https://doi.org/10.1243/09544054JEM990
48. Thrimurthulu, K., Pandey, P., & Nallagundla, V. R. (2004). Optimum part deposition orientation in fused deposition modeling. *International Journal of Machine Tools and Manufacture*, *44*, 585–594. https://doi.org/10.1016/j.ijmachtools.2003.12.004
49. Nancharaiah, T. (2011). Optimization of process parameters in FDM process using design of experiments. *International Journal of Emerging Technologies*, *2*.
50. Gurrala, P., & Regalla, S. (2011). Optimization of support material and build time in fused deposition modeling (FDM). *Applied Mechanics and Materials*, *110–116*. https://doi.org/10.4028/www.scientific.net/AMM.110-116.2245
51. Officer of the Commissioner. (2024, April 22). *U.S. Food and Drug Administration*. FDA. https://www.fda.gov/
52. *ICH Official web site: ICH*. Retrieved April 25, 2024. https://www.ich.org/page/ich-guidelines
53. Quodbach, J., Bogdahn, M., Breitkreutz, J., Chamberlain, R., Eggenreich, K., Elia, A., Gottschalk, N., Gunkel-Grabole, G., Hoffmann, L., Kapote, D., Kipping, T., Klinken, S., Loose, F., Marquetant, T., Windolf, H., Geissler, S., & Spitz, T. (2021). Quality of FDM 3D printed medicines for pediatrics: Considerations for formulation development, filament extrusion, printing process and printer design. *Therapeutic Innovation & Regulatory Science*, *56*. https://doi.org/10.1007/s43441-021-00354-0
54. Scoutaris, N., Ross, S. A., & Douroumis, D. (2018). 3D printed "Starmix" drug loaded dosage forms for paediatric applications. *Pharmaceutical Research*, *35*(2), 34. https://doi.org/10.1007/s11095-017-2284-2
55. Okwuosa, T. C., Stefaniak, D., Arafat, B., Isreb, A., Wan, K.-W., & Alhnan, M. A. (2016). A lower temperature FDM 3D printing for the manufacture of patient-specific immediate release tablets. *Pharmaceutical Research*, *33*(11), 2704–2712. https://doi.org/10.1007/s11095-016-1995-0
56. Goyanes, A., Buanz, A. B. M., Hatton, G. B., Gaisford, S., & Basit, A. W. (2015). 3D printing of modified-release aminosalicylate (4-ASA and 5-ASA) tablets. *European Journal of Pharmaceutics and Biopharmaceutics*, *89*, 157–162. https://doi.org/10.1016/j.ejpb.2014.12.003
57. Goyanes, A., Fina, F., Martorana, A., Sedough, D., Gaisford, S., & Basit, A. W. (2017). Development of modified release 3D printed tablets (printlets) with pharmaceutical excipients using additive manufacturing. *International Journal of Pharmaceutics*, *527*(1), 21–30. https://doi.org/10.1016/j.ijpharm.2017.05.021
58. Khaled, S. A., Alexander, M. R., Wildman, R. D., Wallace, M. J., Sharpe, S., Yoo, J., & Roberts, C. J. (2018). 3D extrusion printing of high drug loading immediate release

paracetamol tablets. *International Journal of Pharmaceutics*, *538*(1), 223–230. https://doi.org/10.1016/j.ijpharm.2018.01.024

59. Gioumouxouzis, C. I., Chatzitaki, A.-T., Karavasili, C., Katsamenis, O. L., Tzetzis, D., Mystiridou, E., Bouropoulos, N., & Fatouros, D. G. (2018). Controlled release of 5-fluorouracil from alginate beads encapsulated in 3D printed ph-responsive solid dosage forms. *AAPS PharmSciTech*, *19*(8), 3362–3375. https://doi.org/10.1208/s12249-018-1084-2
60. Stewart, S. A., Domínguez-Robles, J., McIlorum, V. J., Gonzalez, Z., Utomo, E., Mancuso, E., Lamprou, D. A., Donnelly, R. F., & Larrañeta, E. (2020). Poly(caprolactone)-based coatings on 3D-printed biodegradable implants: A novel strategy to prolong delivery of hydrophilic drugs. *Molecular Pharmaceutics*, *17*(9), 3487–3500. https://doi.org/10.1021/acs.molpharmaceut.0c00515
61. Goyanes, A., Madla, C. M., Umerji, A., Duran Piñeiro, G., Giraldez Montero, J. M., Lamas Diaz, M. J., Gonzalez Barcia, M., Taherali, F., Sánchez-Pintos, P., Couce, M.-L., Gaisford, S., & Basit, A. W. (2019). Automated therapy preparation of isoleucine formulations using 3D printing for the treatment of MSUD: First single-centre, prospective, crossover study in patients. *International Journal of Pharmaceutics*, *567*, 118497. https://doi.org/10.1016/j.ijpharm.2019.118497
62. Alhijjaj, M., Nasereddin, J., Belton, P., & Qi, S. (2019). Impact of processing parameters on the quality of pharmaceutical solid dosage forms produced by fused deposition modeling (FDM). *Pharmaceutics, 11*(12), 633. 10.3390/pharmaceutics11120633. https://doi.org/10.3390/pharmaceutics11120633
63. Azad, M., Olawuni, D., Kimbell, G., Badruddoza, A. Z., Hossain, Md. S., & Sultana, T. (2020). Pharmaceutics polymers for extrusion-based 3D printing of pharmaceuticals: A holistic materials-process perspective. *Pharmaceutics*, *12*(2), 124.
64. Goyanes, A., Buanz, A. B. M., Basit, A. W., & Gaisford, S. (2014). Fused-filament 3D printing (3DP) for fabrication of tablets. *International Journal of Pharmaceutics*, *476*(1–2), 88–92. https://doi.org/10.1016/j.ijpharm.2014.09.044

(e.g., pH, enzymes, redox potential) or external (e.g., temperature, light, magnetic fields) stimuli present in the tumor microenvironment, enabling site-specific drug release and enhanced therapeutic efficacy. This chapter explores the fundamental design principles, mechanisms, and multifunctional characteristics of stimuli-responsive nanocarriers. Emphasis is placed on their role in improving the precision of cancer diagnosis through contrast enhancement and biomarker-targeted imaging, as well as in therapy by facilitating controlled, localized delivery of chemotherapeutic agents. Current advancements in the development of dual- or multi-stimuli-responsive systems and their integration with imaging modalities for theranostic applications are critically reviewed. The chapter concludes by discussing challenges related to biocompatibility, large-scale production, and clinical translation, alongside future prospects in personalized nanomedicine.

3.1 INTRODUCTION TO FUSED DEPOSITION MODELING

Using a common additive manufacturing (AM) method called fusion deposition modeling (FDM), three-dimensional things may be made layer by layer. Stratasys, Scott Crump's business, commercialized it in the 1990s after he conceived it in the late 1980s. The process of FDM involves feeding a thermoplastic filament through a heated extrusion nozzle, which melts the material. After being layered onto a construction platform, the molten plastic rapidly hardens to take on the required shape. The exact motions needed to construct the item are dictated by a CAD file that controls the nozzle. One of FDM's main benefits is its adaptability to a large variety of suitable materials, such as nylon, polyethylene terephthalate glycol (PETG), ABS, and polylactic acid (PLA). This makes it possible to produce items with different mechanical characteristics, hues, and finishes. FDM may be used by both experts and enthusiasts because it is also rather easy to set up and use. FDM is not without its restrictions, though. The layer-by-layer approach may leave obvious layer lines in the final product, which may need to be smoothed down by postprocessing. Furthermore, because the nozzle size and layer height determine the resolution, FDM is not a good option for manufacturing items with fine details or complex geometries [1].

3.1.1 OVERVIEW OF ADDITIVE MANUFACTURING

The last 10 years have seen a significant breakthrough in AM, allowing researchers to create complex, multifunctional, multimaterial designs with

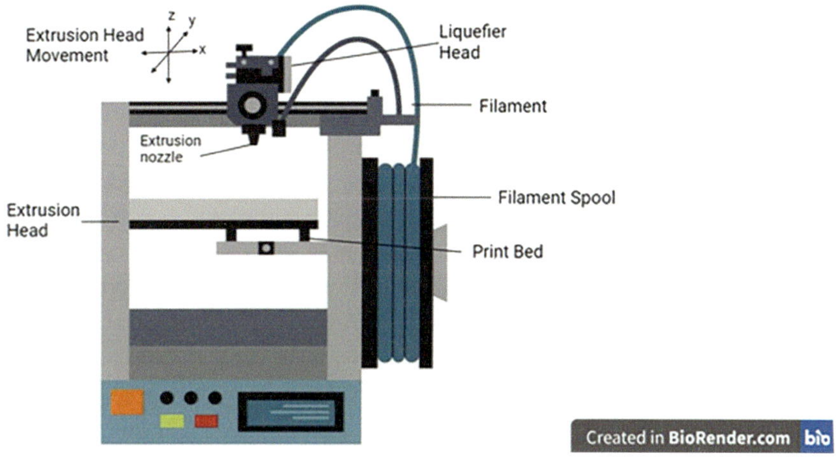

FIGURE 3.1 Fusion deposition modeling (FDM) is a widely used process in 3D printing.

complex geometric properties in a range of settings. AM is not the same as formative and subtractive production as defined by ASTM and ISO standards [2]. Instead, AM is the process of mixing materials to make components using information from 3D models, sometimes layer by layer. Improved designs can be produced, and waste, energy, resource reduction, and supply chain reorganization can all be achieved [3]. AM is the most effective manufacturing method when it comes to situations with high design complexity, low production numbers, and frequent design alterations [4]. Several high-value sectors find AM techniques to be an attractive alternative due to their ability to build geometrically complex components in small batches in an economically viable manner [5].

3.1.2 PRINCIPLES OF FDM TECHNOLOGY

FDM stands as a prevalent AM technique, prized for its affordability and ease of use. Despite rapid technological progress, a major obstacle hindering its widespread adoption remains the challenge of ensuring quality control. Within the AM sphere, particularly in the production of continuous carbon fiber (CF) polymer composites, FDM has garnered significant attention [6]. When contrasted with traditional techniques like autoclave-cured prepreg-based laminates, FDM technology is acknowledged for its capacity to produce continuous fiber-reinforced composites with diminished mechanical properties. Laminates crafted through the FDM process typically exhibit notably inferior tensile, compression, and flexural characteristics compared with those fashioned via conventional manufacturing methodologies [7–13]. One popular 3D printing method for materials based on polymers is fused deposition modeling or FDM. It is the simplest, least expensive, and most widely available approach and belongs to the material extrusion AM technique [14–16]. After extrusion, each layer is cooled and consolidated before being used for further processing. Furthermore, AM may be more environmentally friendly than traditional methods [17]. The fused deposition modeling machine-generated components with the least degree of environmental influence, according to the literature [18]. In practical applications, FDM produces two types of waste materials: support material and prints that are created erroneously. Depending on the printing direction and other variables, the quantity of support material required may change [19].

3.1.3 IMPORTANCE OF MATERIALS SELECTION

Most of the research on fused deposition modeling waste is mostly focused on creating supporting documentation. Put differently, perfect production in optimal conditions is the norm [20]. It is interesting to note that most studies used a variety of industrial and experimental test techniques to examine uniaxial tensile behavior. It is important to emphasize that poor quality and porosity are two common structural issues with FDM printed products that are a result of fused filament-based production [21, 22]. It extends the life of things beyond what was initially planned by the manufacturer. With 3D printing, the element that makes objects obsolete is reduced, and this is not possible with many other technologies [23, 24]. Roughness and manufacturing time optimization have been considered in the FDM technology research. Both conflicting challenges are overcome by applying an adaptive slicing strategy. To evaluate building time and average part surface roughness, several models have been created [25]. Stratasys created fused deposition modeling, or FDM, in Eden Prairie, Minnesota, in the United States. The principal feature of the FDM technology is its capacity to produce prototypes with mechanical and porous properties that may be selectively altered at the local level [26]. The procedure entails using information gleaned from a real item to control the deposit of melted material, which is typically filament shaped. For an FDM system, CAD and CT scan systems typically provide component data in STL format after splitting the model into many slices. This means that when the first layer is printed, the platform moves down so that further layers may be printed on the top of the first one, and this process is repeated. X-Y-Z rectilinear arrangement is the most frequent for the construction. The nozzle may move along the X and Y axes, or horizontal and vertical axes, thanks to a computerized system, and the platform moves along the Z axis [27].

TABLE 3.1 The Significance of Material Choice in 3D Printing.

Aspect	Description	Examples	Reference
Mechanical properties	Different materials exhibit varying mechanical properties such as strength, flexibility, toughness, and impact resistance. Selecting the right material ensures that the printed parts meet the requirements of the intended application	PLA, ABS, TPU	[28]
Thermal properties	Thermal properties like melting point, glass transition temperature, and thermal conductivity affect the material's behavior during printing and its performance in the final application. Choosing materials with suitable thermal properties is essential for achieving successful prints and desired functionality	PETG, Nylon, PC	[28]
Cost	Material cost can significantly impact the overall cost of 3D printing projects, especially for large-scale or high-volume production. Balancing material cost with performance requirements and project budget is essential for cost-effective manufacturing	ASA, PVA, HIPS	[28]

3.2 THERMAL PROPERTIES OF FILAMENT MATERIALS

Filament materials used in 3D printing can have various thermal properties that affect their printability, strength, and application suitability. Melting point temperature at which the filament material changes from a solid to a liquid state, allows it to be extruded through the printer nozzle. For example, PLA typically has a melting point of around 180–220°C, whereas ABS melts at a higher temperature, around 210–250°C. Glass transition temperature is the temperature range at which an amorphous polymer transitions from a hard, glassy state to a rubbery, more flexible state. It affects the material's stiffness and brittleness. For instance, PLA has a Tg of around 60–65°C, whereas ABS has a higher Tg of about 100°C. Heat deflection temperature (HDT) is the temperature at which a material deforms under a specified load. It indicates the material's ability to withstand heat without significant deformation. Materials like ABS typically have higher HDT compared with PLA. Thermal conductivity refers to a material's ability to conduct heat. Higher thermal conductivity can affect printing quality and cooling requirements. Generally, materials like PLA have lower thermal conductivity compared with metals or ceramics. CTE measures how much a material expands or contracts with temperature changes. It is crucial for dimensional accuracy and preventing warping or cracking during printing. Materials with lower CTE are less prone to deformation. Thermal stability refers to a material's ability to maintain its properties at elevated temperatures without degradation. Some materials may experience degradation or decomposition when subjected to high temperatures over extended periods, affecting print quality and durability [29].

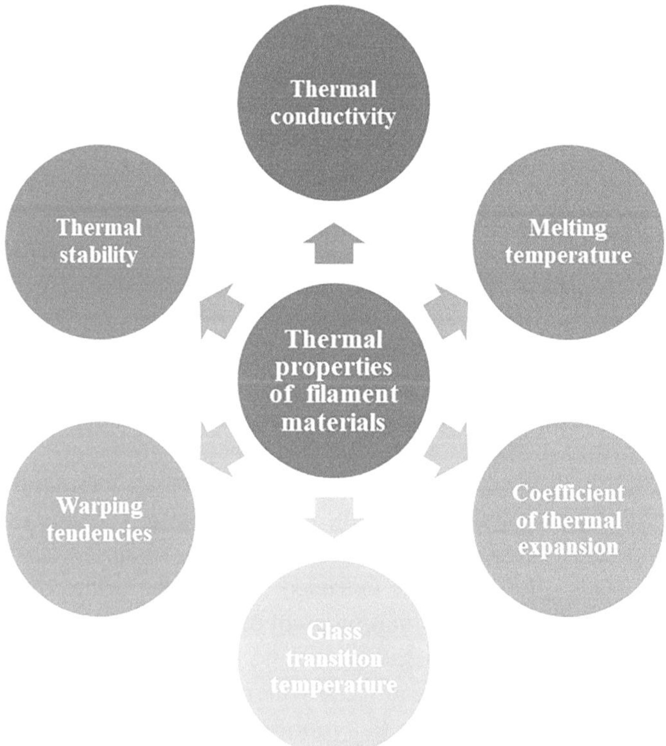

FIGURE 3.2 Common FDM filament material properties.

3.2.1 MELT FLOW CHARACTERISTICS

The manufacture and model of the printer are important choices since machine-specific factors relate to the special features of the printer itself. The printing settings define parameters specific to each task, whereas the characteristics of the filament determine material-specific parameters. Printers enable adjustment of factors intrinsic to their operation, like temperature and print speed. Nevertheless, the rheology required for melting and feeding during FDM printing is absent from most medicinal polymers. To achieve optimal printing outcomes, it is crucial to accurately adjust processing parameters like printing resolution, infill density, layer height, extrusion speed, nozzle diameter, temperature, and bed temperature. Although several materials are used in 3D printing, the most commonly used ones include HDPE, ABN, PLA, Nylon, and PC. The differences in these materials' melting points have a significant impact on the final products' characteristics. In the print head, filaments melt during printing. There is a 150–250°C temperature range, depending on the technology being used. The optimum temperature is chosen for the cooling substrate that is used for printing to provide uniform cooling and deformation (40–100)°C is the usual range [30].

3.2.2 THERMAL STABILITY

Filament serves as the core material in FDM printing, typically comprising pure, low-melting-point polymers. However, there is a need to enhance the strength of these polymers at times. To address this, various researchers and companies have devised polymer composite filaments. These filaments amalgamate the base polymer with reinforcing elements, resulting in 3D printing materials that offer structural integrity and functional advantages beyond what standard polymers can achieve [31]. FDM material may be directly made from pure polymer-based decorations, which are frequently offered for sale. With each reinforcement in a composite polymer offering its own distinct characteristics, it is crucial to handle the process of making composite filaments with precision and care [32–35]. Polymer filament used in FDM manufacturing can be created through the extrusion process using either polymer pellets or raw polymer materials. In this process, extruders are employed to exert pressure or force on the material, pushing it through die-holes to form an extrudate [36]. Crucial FDM process parameters might be related to the material, the machine, or

infill density affect printed goods, Samykano et al. [55] employed FDM technology and ABS material. The optimum parameters for achieving the highest mechanical strength were identified as a result of this analysis. Chaudhry and Czekanski's findings suggest that the primary factor influencing the tensile strength of a 3D sample created through FDM technology is the layer height, whereas the infill percentage also plays a significant role in determining its mechanical strength. Additionally, their research establishes a correlation between various printing parameters and the mechanical properties exhibited by 3D printed samples [56].

3.3.2 MODULUS OF ELASTICITY

The mechanical properties of components produced via FDM are significantly influenced by factors such as layer thickness, raster width, fill percentage, raster angle, and build orientation [57–60]. Using a range of optimization techniques, prior research has optimized these process parameters to improve the mechanical properties of the component. Based on experimental results, Ahn et al. [61] created building recommendations for FDM components by analyzing the anisotropic material properties of FDM using ABS. To extend the applications of FDM, the variety of FDM materials is constantly growing. The selection of materials for FDM is typically confined to thermoplastics and their blends due to the operating limitations of most commercial FDM extruders. These machines are typically designed to work with materials that can solidify effectively at temperatures of 300°C or higher, preventing issues like warping in the printed layers. High glass transition temperatures are found in some materials, such as PEEK. Unfortunately, unless they replace the thermoset with a thermocouple and alter the extrusion head, most commercial FDM systems are unable to reach the extrusion temperature required by PEEK, which is above 300°C [62].

3.3.3 IMPACT RESISTANCE

Adjusting various printing parameters such as layer thickness, orientation of layers, shell count, nozzle-to-bed distance, and so forth, can alter the mechanical properties during the FDM process. Numerous scholars have studied the aspects of printing and how they impact final product attributes like surface roughness. Due to the melting and solidification of the material, residual stresses are produced throughout the production process. These stresses are brought on by temperature gradients and contractions that take place during the process. Kantaros and Karalekas found that employing a longitudinal (0°) orientation and a layer thickness of 0.25 mm resulted in minimal residual strain [63]. They looked at how the residual strain was affected by the printing direction and layer thickness. Moreover, the anticipated variance in mechanical properties between printed parts and standard ABS material stems from research suggesting that the adhesion resistance of the filaments is inferior to that of ABS [64].

3.3.4 DIMENSIONAL ACCURACY

The research authors produced a total of 18 × 3 ASTM D683 specimens using various processing procedures. A batch of three copies of each specimen was created in order to confirm the repeatability of the FDM approach for the same set of processing parameters [65]. To explore the impact of various manufacturing parameters on dimensional accuracy, this study involved the production of eight ASTM D638 type IV tensile test specimens. Four specimens were manufactured with different extrusion temperatures while keeping other processing parameters constant. The remaining four specimens were printed with varying layer thicknesses while maintaining consistent processing conditions. Measurements were taken at designated points on each specimen, and the data were compared with the input from the CAD model [66].

3.4 COMMONLY USED FDM FILAMENT MATERIALS

Commonly used FDM filament materials include PLA, known for its biodegradability and ease of printing, suitable for prototyping and hobbyist projects. ABS offers durability and higher temperature resistance, ideal for functional prototypes and mechanical parts. PETG is strong, with minimal warping, suitable for mechanical parts and food-safe containers. Thermoplastic polyurethane (TPU) provides flexibility and chemical resistance, often used for flexible parts and seals. Nylon offers toughness and high tensile strength, suitable for gear and wearable items. ASA is weather-resistant, and ideal for outdoor applications, whereas PC offers high strength and transparency for engineering parts and enclosures. HIPS serves as support material for ABS prints and packaging [67].

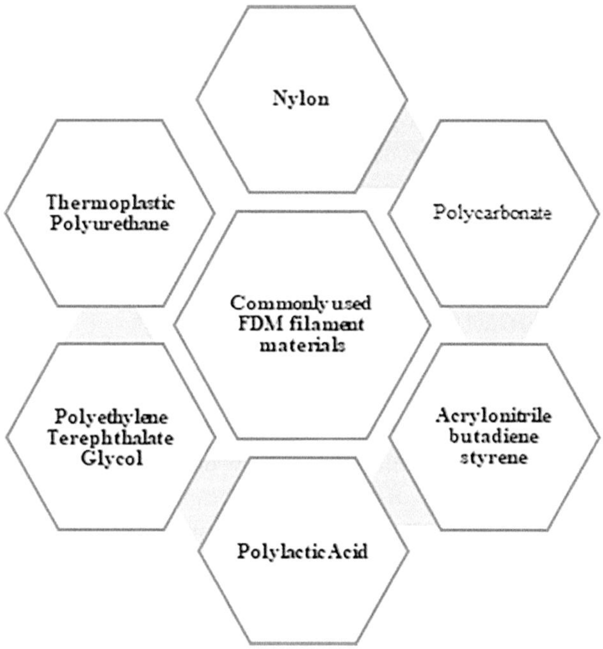

FIGURE 3.3 Most popular flexible 3D printing technique with versatile filament compatibility.

3.4.1 ACRYLONITRILE BUTADIENE STYRENE

ABS is an amorphous polymer that is thermoplastic and commonly used in the FDM process to print items three-dimensionally. ABS is a thermoplastic with broad applications across several industries because of its strong, flexible, resilient, production-grade, chemical resistance, and relatively low water absorption qualities. Furthermore, to ensure good color fastness, antioxidants significantly improve oxidation stability. Polylactic acid (PLA) melts at a lower temperature than ABS, which typically has an amorphous melting point of 230°. Though it has a lower likelihood of clogging a nozzle than PLA, ABS is not biodegradable [68]. Applications include protective helmets for sports equipment, pipes and accessories, medical equipment, hub caps, body panels, instrument panels, vacuum cleaners, furnishings, hoses, grommets, and household appliances [69, 70].

3.4.2 POLYLACTIC ACID

PLA is a thermoplastic polymer derived mainly from maize starch, and is recognized for its biodegradability, biocompatibility, and renewable nature. Its production is straightforward, and it exhibits impressive mechanical characteristics. It is capable of substituting petroleum-based polymers. Furthermore, pieces have been fabricated using FDM using PLA-ISO and PLA-ABS as filaments [71]. The hydrophobic polymer PLA, on the other hand, has low toughness, a slow rate of breakdown, and limited heat stability [72]. With a melting point ranging from 170 to 180°C, PLA is a homopolymer [73]. Products created from PLA include plastic films, bottles, biodegradable medical equipment, and scaffolds [74].

3.4.3 POLYETHYLENE TEREPHTHALATE GLYCOL (PETG)

PETG is a co-polyester composed of thermoplastic plastic that combines chemical resistance, high mechanical properties, and ease of usage for 3D printing. PETG is extensively utilized across various applications because of its remarkable strength, durability, and ability to withstand damage. It is widely utilized to produce completed products, functional prototypes, and mechanical parts. PETG's robust tensile strength allows its mechanical properties to withstand weights and forces. It also has outstanding impact resistance, so it will not break or shatter in the event of unanticipated forces or collisions. Process variables' effects on the mechanical properties of PETG materials have been examined [75–83].

3.4.4 THERMOPLASTIC POLYURETHANE (TPU)

TPU is a commonly used polyurethane (PU) material. It is made by using hard and soft polymeric segments in different combinations. Soft segments with a low glass Tg are necessary to create continuous matrices in polymer nanocomposites with significant flexibility at low temperatures, whereas hard segments are more likely to self-assemble into domains by a crosslinking action [84]. All things considered, physical crosslinks seem to be reversible, suggesting that the hard and soft segments share the soft segment's properties to produce a homogeneous mixture that goes beyond Tg [85]. This distinctive quality is contingent upon the processing conditions of the material as well as the morphological characteristics linked to certain chemical components [86]. This has led to the widespread usage of TPU in many different applications, such as scaffolds, biomaterials, and coatings [87].

3.5 FUTURE PROSPECTS

The choice of materials for FDM is changing, extending the possibilities and uses of this technique for AM [88]. Research and development activities are expanding the variety of materials suitable for FDM, which is influencing its future. Previously, the industry relied solely on thermoplastics like PLA and ABS [89]. The development of engineering-grade thermoplastics is being fueled by the need for FDM-printed parts in functional prototypes and end-use applications. Materials that offer improved mechanical characteristics, heat resistance, and chemical resistance include polycarbonate (PC), polyetherimide (PEI), and PETG. Due to their strict performance requirements, the automotive, aerospace, and healthcare industries are among those that FDM can now serve [90]. High-performance composite components are made with FDM when reinforcing fibers like CF, fiberglass, or Kevlar are integrated into thermoplastic matrices [91–95]. These composite materials have improved stiffness, strength, and endurance, which makes them appropriate for demanding uses in the sporting goods, aerospace, and defense industries. There is a drive for FDM to use recycled and bio-based materials to address environmental issues. These materials, which are made from recycled polymers or renewable resources, have mechanical qualities that are on par with those of conventional thermoplastics but also have advantages for sustainability [96]. They appeal to companies looking to lessen their environmental impact as well as customers who care about the environment. Parts with specialized qualities like conductivity, flame resistance, or antibacterial qualities may be produced using FDM filaments by including functional additives [97]. In industries where certain material characteristics are essential, such as electronics, protective gear, and healthcare, this creates prospects for FDM [98]. The ability to print several materials at once is being made possible by advances in FDM technology, which enable the creation of intricate assemblies with several components in a single build process [99]. This feature opens up new options for product development and customization by increasing the design freedom and functionality of

FDM-printed parts [100]. In conclusion, the continuous development of materials compatible with FDM from engineering-grade thermoplastics to bio-based materials and functional additives along with the progress made in multimaterial printing, places FDM in a position where it is a flexible and increasingly important technology for a variety of industries [101].

3.6 CONCLUSION

A crucial factor that greatly affects the functionality, characteristics, and use of 3D-printed components is the choice of materials for FDM. In this chapter, we have explored various factors that play a pivotal role in material selection for FDM, encompassing thermal properties, mechanical characteristics, chemical resistance, environmental considerations, ease of printing, color options, cost implications, and the availability of specialty filaments. Here, we summarize the key insights and conclusions drawn from our exploration. First, the thermal properties of materials are crucial for FDM, as they dictate the extrusion and layer adhesion processes during printing. Thermoplastics like ABS, PLA, PETG, and Nylon are commonly chosen for their favorable melting points and thermal stability, enabling efficient printing without compromising part integrity. Second, mechanical properties such as tensile strength, impact resistance, flexibility, and hardness are paramount considerations in material selection. Engineers must assess these properties to ensure that the printed parts meet the performance requirements of their intended applications. For instance, ABS offers good overall mechanical properties, whereas PLA is known for its stiffness and ease of printing. Third, the chemical resistance of materials is essential for applications exposed to harsh environments or chemical substances. Materials like PETG exhibit superior chemical resistance compared with others, making them suitable for such demanding scenarios. Fourth, environmental factors such as moisture absorption, UV resistance, and biodegradability should be considered, particularly for outdoor or eco-friendly applications. PLA, with its biodegradability and renewable sourcing, stands out as an environmentally friendly option. Fifth, ease of printing is a critical aspect that affects the overall success of FDM projects. Materials should exhibit good printability characteristics, including minimal warping, low shrinkage, and excellent adhesion between layers. PLA is favored by many due to its ease of printing and minimal odor. Moreover, the color options available for FDM materials provide opportunities for customization and aesthetic appeal, catering to a wide range of design preferences and applications. Cost considerations play a significant role, especially for large-scale production or budget-constrained projects. While PLA is generally more affordable, specialty filaments like TPU or metal composites offer unique properties but may come with higher costs. Last, the availability of specialty filaments opens possibilities for novel applications, including conductive materials, wood-filled PLA, and flexible TPE. These materials offer unique properties that expand the potential applications of FDM technology.

COMPETING INTEREST STATEMENT

The authors declare no conflict of interest.

DISCLAIMER

None

KEYWORDS

- **additive manufacturing**
- **fused deposition modeling**
- **materials selection**
- **mechanical properties**
- **thermal properties**

REFERENCES

1. Li, Q., Huang, T., Liu, J., & Tan, L. (2024). Time-series vision transformer based on cross space-time attention for fault diagnosis in fused deposition modelling with reconstruction of layer-wise data. *Journal of Manufacturing Processes, 115*, 240–255.
2. Pérez, M., Carou, D., Rubio, E. M., & Teti, R. (2020). Current advances in additive manufacturing. *Procedia CIRP*, *88*, 439–444.
3. Peng, T., Kellens, K., Tang, R., Chen, C., & Chen, G. (2018). Sustainability of additive manufacturing: An overview on its energy demand and environmental impact. *Additive Manufacturing, 21*, 694–704.

4. Abdulhameed, O., Al-Ahmari, A., Ameen, W., & Mian, S. H. (2019). Additive manufacturing: Challenges, trends, and applications. *Advances in Mechanical Engineering, 11(2)*, 1687814018822880.
5. Snow, Z., Diehl, B., Reutzel, E. W., & Nassar, A. (2021). Toward in-situ flaw detection in laser powder bed fusion additive manufacturing through layerwise imagery and machine learning. *Journal of Manufacturing Systems, 59*, 12–26.
6. Galos, J., & Wang, X. (2021). Demonstration of computer vision for void characterisation of 3D-printed continuous carbon fibre composites. *Results Materials, 22*, 100566.
7. Dickson, A. N., Barry, J. N., McDonnell, K. A., & Dowling, D. P. (2017). Fabrication of continuous carbon, glass and Kevlar fibre reinforced polymer composites using additive manufacturing. *Additive Manufacturing, 16*, 146–152.
8. Justo, J., Távara, L., García-Guzmán, L., & París, F. (2018). Characterization of 3D printed long fibre reinforced composites. *Composite Structures, 185*, 537–548.
9. Blok, L. G., Longana, M. L., Yu, H., & Woods, B. K. S. (2018). An investigation into 3D printing of fibre reinforced thermoplastic composites. *Additive Manufacturing, 22*, 176–186.
10. Dul, S., Fambri, L., & Pegoretti, A. (2016). Fused deposition modelling with ABS–graphene nanocomposites. *Composite Part A Applied Science and Manufacturing, 85*, 181–191.
11. Araya-Calvo, M., López-Gómez, I., Chamberlain-Simon, N., León-Salazar, J. L., Guillén-Girón, T., Corrales-Cordero, J. S., et al. (2018). Evaluation of compressive and flexural properties of continuous fiber fabrication additive manufacturing technology. *Additive Manufacturing, 22*, 157–164.
12. Caminero, M. A., Chacón, J. M., García-Moreno, I., & Rodríguez, G. P. (2018). Impact damage resistance of 3D printed continuous fibre reinforced thermoplastic composites using fused deposition modelling. *Composite Part B Engineering, 148*, 93–103.
13. Goh, G. D., Dikshit, V., Nagalingam, A. P., Goh, G. L., Agarwala, S., Sing, S. L., et al. (2018). Characterization of mechanical properties and fracture mode of additively manufactured carbon fiber and glass fiber reinforced thermoplastics. *Materials and Design, 137*, 79–89
14. Cuiffo, M. A., Snyder, J., Elliott, A. M., Romero, N., Kannan, S., & Halada, G. P. (2017). Impact of the fused deposition (FDM) printing process on polylactic acid (PLA) chemistry and structure. *Applied Science, 7*(6), 579.
15. Dey, A., & Yodo, N. (2019). A systematic survey of FDM process parameter optimization and their influence on part characteristics. *Journal of Manufacturing Materials, 3*(3), 64.
16. Valino, A. D., Dizon, J. R. C., Espera, A. H., Chen, Q., Messman, J., & Advincula, R. C. (2019). Advances in 3D printing of thermoplastic polymer composites and nanocomposites. *Progress in Polymer Science, 98*, 101162.
17. Mustapha, K. B., & Metwalli, K. M. (2021). A review of fused deposition modelling for 3D printing of smart polymeric materials and composites. *European Polymer Journal, 156*, 110591.
18. Madhu, N. R., Erfani, H., Jadoun, S., Amir, M., Thiagarajan, Y., & Chauhan, N. P. S. (2022). Fused deposition modelling approach using 3D printing and recycled

industrial materials for a sustainable environment: A review. *The International Journal of Advanced Manufacturing Technology, 122*(5–6), 2125–2138.

19. Zhang, X., Fan, W., & Liu T. (2020). Fused deposition modeling 3D printing of polyamide-based composites and its applications. *Composites CommUnications, 21*, 100413.
20. Dumpa, N., Butreddy, A., Wang, H., Komanduri, N., Bandari, S., & Repka, M. A. (2021). 3D printing in personalized drug delivery: An overview of hot-melt extrusion-based fused deposition modeling. *International Journal of Pharmaceutics, 600*, 120501.
21. Gordeev, E. G., Galushko, A. S., & Ananikov, V. P. (2018). Improvement of quality of 3D printed objects by elimination of microscopic structural defects in fused deposition modeling. *PLoS One, 13*(6), e0198370.
22. Galantucci, L. M., Lavecchia, F., & Percoco, G. (2009). Experimental study aiming to enhance the surface finish of fused deposition modeled parts. *CIRP Annals, 58*(1), 189–192.
23. Santana, A., Afonso, P., Zanin, A., & Wernke, R. (2017). Costing models for capacity optimization in Industry 4.0: Trade-off between used capacity and operational efficiency. *Procedia Manufacturing, 13*, 1183–1190.
24. Bourell, D. L., Rosen, D. W., & Leu, M. C. (2014). The roadmap for additive manufacturing and its impact. *3D Printing and Additive Manuf*acturing, *1*(1), 6–9.
25. Thrimurthulu, K., Pandey, P. M., & Reddy, N. V. (2004). Optimum part deposition orientation in fused deposition modeling. *International Journal of Machine Tools and Manufacture, 44*(6), 585–594.
26. Sun, Q., Rizvi, G. M., Bellehumeur, C. T., & Gu, P. (2008). Effect of processing conditions on the bonding quality of FDM polymer filaments. *Rapid Prototyping Journal, 14*(2), 72–80.
27. Wiria, F. E., Sudarmadji, N., Leong, K. F., Chua, C. K., Chng, E. W., & Chan, C. C. (2020). Selective laser sintering adaptation tools for cost effective fabrication of biomedical prototypes. *Rapid Prototyping Journal, 16*(2), 90–99.
28. Mohd Puad, N. A. S., Abdul Haq, R. H., Mohd Noh, H., Abdullah, H. Z., Idris, M. I., & Lee, T. C. (2020). Review on the fabrication of fused deposition modelling (FDM) composite filament for biomedical applications. *Materials Today Proceedings, 29*, 228–232.
29. Araújo, M. R. P., Sa-Barreto, L. L., Gratieri, T., Gelfuso, G. M., & Cunha-Filho, M. (2019). The digital pharmacies era: How 3D printing technology using fused deposition modeling can become a reality. *Pharmaceutics, 11*(3), 128.
30. Trhlíková, L., Zmeskal, O., Psencik, P., & Florian, P. (2016). Study of the thermal properties of filaments for 3D printing. *AIP Conference Proceedings,* 1752.
31. Wang, X., Jiang, M., Zhou, Z., Gou, J., & Hui, D. (2017). 3D printing of polymer matrix composites: A review and prospective. *Composites Part B Engineering, 110*, 442–458.
32. Milosevic, M., Stoof, D., & Pickering, K. L. (2017). Characterizing the mechanical properties of fused deposition modelling natural fiber recycled polypropylene composites. *Journal of Composite Science, 1*(1), 7.
33. Yetgin, S. H. (2020). Tribological properties of compatabilizer and graphene oxide-filled polypropylene nanocomposites. *Bulletin of Materials Science, 43*(1), 1–8.

34. Nancharaiah, T., Raju, R., & Raju, R. (2010). An experimental investigation on surface quality and dimensional accuracy of FDM components. *International of Journal of Emerging Technology, 1*(2), 106–111.
35. Shabdin, M. K., Rahman, M. A. A., Mazlan, S. A., Ubaidillah, A., Hapipi, N. M., Adiputra, D., et al. (2019). Material characterizations of Gr-based magnetorheological elastomer for possible sensor applications: Rheological and resistivity properties. *Material, 12*(3), 391.
36. Rauwendaal, C. (2014). Polymer Extrusion. *Polymer Extrusion*, 1–10
37. Nasereddin, J. M., Wellner, N., Alhijjaj, M., Belton, P., & Qi, S. (2018). Development of a simple mechanical screening method for predicting the feedability of a pharmaceutical FDM 3D printing filament. *Pharmaceutical Research, 35*, 1–3.
38. Ursan, I., Chiu, L., & Pierce, A. (2013). Three-dimensional drug printing: A structured review. *Journal of the American Pharmacists Association, 53*(2), 136–144.
39. Goyanes, A., Det-Amornrat, U., Wang, J., Basit, A. W., & Gaisford, S. (2016). 3D scanning and 3D printing as innovative technologies for fabricating personalized topical drug delivery systems. *Journal of Controlled Release, 234*, 41–48.
40. Norman, J., Madurawe, R. D., Moore, C. M. V., Khan, M. A., & Khairuzzaman, A. (2017). A new chapter in pharmaceutical manufacturing: 3D-printed drug products. *Advanced Drug Delivery Reviews, 108*, 39–50.
41. Gao, G., Xu, F., Xu, J., Tang, G., & Liu, Z. (2022). A survey of the influence of process parameters on mechanical properties of fused deposition modeling parts. *Micromachines, 13*(4), 553.
42. Syrlybayev, D., Zharylkassyn, B., Seisekulova, A., Perveen A., Talamona D., Bruno C., et al. (2021). Optimization of the warpage of fused deposition modeling parts using finite element method. *Polymers, 13*(21), 3849.
43. Huynh, T. T., Tien, V. T., Nguyen, Q. M., & Nguyen, T. K. (2021). Minimizing warpage for macro-size fused deposition modeling parts. *Computers Materials and Continua, 68*(3), 2913–2923.
44. Peng, A. H. (2012). Research on the interlayer stress and warpage deformation in FDM. *Advanced Materials Research, 538*, 1564–1567.
45. Armillotta, A., Bellotti, M., & Cavallaro M. (2018). Warpage of FDM parts: Experimental tests and analytic model. *Robot and Computer Integrated Manufacturing, 50*, 140–152.
46. De Macedo, R. Q., & Ferreira, R. T. L. (2017). December. COBEM-2017-0124 residual thermal stress in fused deposition modelling. In *Proceedings of the 24th ABCM International Congress of Mechanical Engineering, Curitiba, Brazil* (pp. 3–8).
47. Zhang, Y., & Chou, Y. (2006). Three-dimensional finite element analysis simulations of the fused deposition modelling process. *Proceedings of the Institution of Mechanical Engineers Part B Journal of Engineering Manufacture, 220*(10), 1663–1671.
48. Chockalingam, K., Karthic, M., & Jawahar, N. (2018). Development of thermal and structural deformation model to predict the part build dimensional error in fused deposition modeling. *ADMT Journal, 11*, 27–37.
49. Yadav, D., Chhabra, D., Garg, R. K., Ahlawat, A., & Phogat, A. (2020). Optimization of FDM 3D printing process parameters for multi-material using artificial neural network. *Materials Today Proceedings, 21*, 1583–1591.

50. Goudswaard, M., Hicks, B., & Nassehi, A. (2021). The creation of a neural network based capability profile to enable generative design and the manufacture of functional FDM parts. *The International Journal of Advanced Manufacturing Technology, 113*, 2951–2968.
51. Mahapatra, S. S., & Sood, A. K. (2021). Bayesian regularization-based Levenberg-Marquardt neural model combined with BFOA for improving surface finish of FDM processed part. *The International Journal of Advanced Manufacturing Technology, 12*, 1223–1235.
52. Zhao, Y., Chen, Y., & Zhou, Y. (2019). Novel mechanical models of tensile strength and elastic property of FDM AM PLA materials: Experimental and theoretical analyses. *Material Design, 5*,181.
53. Wang, P., Zou, B., Xiao, H., Ding, S., & Huang, C. (2019). Effects of printing parameters of fused deposition modeling on mechanical properties, surface quality, and microstructure of PEEK. *Journal of Materials Processing Technology, 271*, 62–74.
54. Wang, P., Zou, B., Ding, S., Li, L., & Huang, C. (2021). Effects of FDM-3D printing parameters on mechanical properties and microstructure of CF/PEEK and GF/PEEK. *Chinese Journal of Aeronautics, 9*, 236–246.
55. Samykano, M., Selvamani, S. K., Kadirgama, K., Ngui, W. K., Kanagaraj, G., & Sudhakar, K. (2019). Mechanical property of FDM printed ABS: Influence of printing parameters. *The International Journal of Advanced Manufacturing Technology, 102*, 79–96.
56. Chaudhry, M. S., & Czekanski, A. (2020). Evaluating FDM process parameter sensitive mechanical performance of elastomers at various strain rates of loading. *Materials, 13*, 3202.
57. Sood, A. K., Ohdar, R. K., & Mahapatra, S. S. (2010). Parametric appraisal of mechanical property of fused deposition modelling processed parts. *Materials and Design, 28*, 7–9.
58. Lee, B. H., Abdullah, J., Khan, Z. A. (2005). Optimization of rapid prototyping parameters for production of flexible ABS object. *Journal of Materials Processing Technology, 169*, 54–61.
59. Sun, Q., Rizvi, G. M., Bellehumeur, C. T., & Gu, P. (2008). Effect of processing conditions on the bonding quality of FDM polymer filaments. *Rapid Prototyping Journal, 14*(2), 72–80.
60. Wu,W., Geng, P., Li, G., Zhao, D., Zhang, H., & Zhao, J. (2015). Influence of layer thickness and raster angle on the mechanical properties of 3D-printed PEEK and a comparative mechanical study between PEEK and ABS. *Materials, 8*, 5834–5846.
61. Ahn, S. H., Montero, M., Odell, D., Roundy, S.; & Wright, P. K. (2002). Anisotropic material properties of fused deposition modeling ABS. *Rapid Prototyping Journal, 2*, 48–57.
62. Garg, A., & Bhattacharya, A. (2017). An insight to the failure of FDM parts under tensile loading: finite element analysis and experimental study. *International Journal of Mechanical Science, 120*, 25–36.
63. Kantaros, A., & Karalekas, D. (2013). Fiber Bragg grating based investigation of residual strains in ABS parts fabricated by fused deposition modeling process. *Materials and Design 50*, 44–50.

64. Rodríguez, J. F., Thomas, J. P., & Renaud, J. E. (2003). Mechanical behavior of acrylonitrile butadiene styrene fused deposition materials modeling. *Rapid Prototyping Jounal, 4*, 219–230.
65. Lafaghani, A., Qattawi, A., Alrawi, B., & Guzman, A. (2017). Experimental optimization of fused deposition modelling processing parameters: A design-for-manufacturing approach. *Procedia Manufing, 10*, 791–803.
66. Alafaghani, A., Qattawi, A., & Ablat, M. A. (2017). Design consideration for additive manufacturing: Fused deposition modelling. *Open Journal of Applied Science, 6*, 291–318.
67. Brebu, M., Bhaskar, T., Murai, K., Muto, A., Sakata, Y., & Uddin, M. A. (2004). The individual and cumulative effect of brominated flame retardant and polyvinylchloride (PVC) on thermal degradation of acrylonitrile–butadiene–styrene (ABS) copolymer. *Chemosphere, 56*, 33–40.
68. Dey, A., & Yodo, N. (2019). A systematic survey of FDM process parameter optimization and their influence on part characteristics. *Journal of Manufacturing and MaterIals, 3*, 64.
69. Begum, S. A., Rane, A. V., & Kanny, K. (2020). Applications of compatibilized polymer blends in automobile industry. In *Compatibilization of polymer blends: Micro and nano scale phase morphologies, interphase characterization, and properties* (Vol. 5, pp, 63–93).
70. N, J., & Senthil, P. (2019). Application of 3D printed ABS based conductive carbon black composite sensor in void fraction measurement. *Composites Part B Engineering, 159*, 24–30.
71. Deshmukh, K., Ahamed, B., Deshmukh, R. R., Khadheer Pasha, S. K., Bhagat, P. R., Chidambaram K. (2017). Biopolymer composites with high dielectric performance: Interface engineering. In *Biopolymer composites in electronics* (pp. 27–128).
72. Patra, A. K., & Pattanayak, A. K. (2015). Novel varieties of denim fabrics. In *Denim: Manufacture, finishing and applications* (Vol. *23*, pp. 483–506).
73. Garlotta, D. (2001). A literature review of poly(Lactic Acid). *Journal of Environmental Polymers and Degradation, 9*, 2.
74. DeStefano, V., Khan, S., & Tabada, A. (2020). Applications of PLA in modern medicine. *Engineered Regeneration, 1*, 76–87.
75. Hsueh, M. H., Lai, C. J., Wang, S. H., Zeng, Y. S., Hsieh, C. H., & Pan, C. Y. (2021). Effect of printing parameters on the thermal and mechanical properties of 3D-printed PLA and PETG, using fused deposition modeling. *Polymers, 27*, 1758.
76. Dizon, J. R. C., Gache, C. C. L., Cascolan, H. M. S., Cancino, L. T., Advincula, R. C. (2021). Post-processing of 3D-printed polymers. *Technologies, 9*, 61.
77. Panneerselvam, T., Raghuraman, S., & Krishnan, V. (2021). Investigating mechanical properties of 3d-printed polyethylene terephthalate glycol material under fused deposition modeling. *Journal of the Institution of Engineers (India): Series C, 102*, 75–87.
78. Mercado-Colmenero, J. M., & MartinDoñate, C. (2020). Experimental and numerical analysis for the mechanical characterization of PETG polymers manufactured with FDM technology under pure uniaxial compression stress states for architectural applications. *Polymers (Basel), 12*, 1–25.

79. Barrios, J. M., & Romero, P. E. (2019). Improvement of surface roughness and hydrophobicity in PETG parts manufactured via fused deposition modeling (FDM): An application in 3d printed self–cleaning parts. *Materials, 12*, 2499.
80. Valvez, S., Silva, A. P., & Reis, P. N.B. (2022). Compressive behaviour of 3D-printed PETG composites. *Aerospace*, *9*, 124.
81. Kasmi, S., Ginoux, G., Allaoui, S., & Alix, S. (2021). Investigation of 3D printing strategy on the mechanical performance of coextruded continuous carbon fiber reinforced PETG. *Journal of Applied Polymers Science, 138*, 50955.
82. Kováčová, M., Kozakovičová, J., & Procházka, M. (2020). Novel hybrid PETG composites for 3D printing. *Applied Science, 10*, 3062.
83. Kumar, K. S., Soundararajan, R., Shanthosh, G., Saravanakumar, P., & Ratteesh, M. (2021). Augmenting effect of infill density and annealing on mechanical properties of PETG and CFPETG composites fabricated by FDM. *Materials Today Proceedings, 21*, 86–91.
84. He, Y., Xie, D., & Zhang, X. (2014). The structure, microphase-separated morphology, and property of polyurethanes and polyureas. *Journal of Materials Science, 73*, 39–52.
85. Driffield, M., Bradley, E. L., & Castle, L. (2007). A method of test for residual isophorone diisocyanate trimer in new polyester-polyurethane coatings on light metal packaging using liquid chromatography with tandem mass spectrometric detection. *Journal of Chromatography A, 6*, 1–6.
86. Hernandez, R., Weksler, J., Padsalgikar, A., & Runt J. (2008). In vitro oxidation of high polydimethylsiloxane content biomedical polyurethanes: Correlation with the microstructure. *Journal of Biomedical Materials Research Part A*, *87*, 46–56.
87. Chen, Q., Mangadlao, J. D., Wallat, J., & De Leon, A. (2017). 3D printing biocompatible polyurethane/poly(lactic acid)/graphene oxide nanocomposites: Anisotropic properties. *ACS Applied Materials Interfaces*, *40*, 15–23.
88. Rajan, K., Samykano, M., Kadirgama, K., Harun, W. S., & Rahman, M. M. (2022). Fused deposition modeling: Process, materials, parameters, properties, and applications. The *International Journal of Advanced Manufacturing Technology, 15*, 31–70.
89. Vyavahare, S., Teraiya, S., Panghal, D., & Kumar, S. (2020). Fused deposition modelling: A review. *Rapid Prototyping Journal, 26*, 176–201.
90. Mwema, F. M., Akinlabi, E. T., Mwema, F. M., & Akinlabi, E. T. (2020). Basics of fused deposition modelling (FDM). In *Fused deposition modeling: Strategies for quality enhancement* (pp. 1–5).
91. Shanmugam, V., Pavan, M. V., Babu, K., & Karnan, B. (2021). Fused deposition modeling based polymeric materials and their performance: A review. *Polymers Composites, 42*, 56–77.
92. Krajangsawasdi, N., Blok, L. G., Hamerton, I., Longana, M. L., Woods, B. K., & Ivanov, D. S. (2021). Fused deposition modelling of fibre reinforced polymer composites: A parametric review. *Journal of Composites Science, 16*, 29.
93. Daminabo, S. C., Goel, S., Grammatiko, S. A., Nezhad, H. Y., Thakur, V. K. (2020). Fused deposition modeling-based additive manufacturing (3D printing): Techniques for polymer material systems. *Materials Today Chemistry, 16*, 100248.

94. BacaLopez, D. M., & Ahmad, R. (2020). Tensile mechanical behaviour of multi-polymer sandwich structures via fused deposition modelling. *Polymers, 12*, 651.
95. Dhinakaran, V., Kumar, K. M., & Ram, P. B (2020). A review on recent advancements in fused deposition modeling. *Materials Today Proceedings, 75*, 2–6.
96. Sheoran, A. J., & Kumar, H. (2020). Fused deposition modeling process parameters optimization and effect on mechanical properties and part quality: Review and reflection on present research. *Materials Today Proceedings, 16*, 59–72.
97. Bakır, A. A., Atik, R., & Özerinç, S. (2021). Mechanical properties of thermoplastic parts produced by fused deposition modeling: a review. *Rapid Prototyping Journal, 27*, 37–61.
98. Mustapha, K. B., & Metwalli, K. M. (2021). A review of fused deposition modelling for 3D printing of smart polymeric materials and composites. *European Polymers Journal, 156*, 110591.
99. Pu'ad, N. M., Haq, R. A., Noh, H. M., & Abdullah, H. Z. (2020). Review on the fabrication of fused deposition modelling (FDM) composite filament for biomedical applications. *Materials Today Proceeding, 29*, 28–32.
100. Naveed, N. (2021). Investigate the effects of process parameters on material properties and microstructural changes of 3D-printed specimens using fused deposition modelling (FDM). *Advanced Materials Technology, 36*, 17–30.
101. Penumakala, P. K., Santo, J., & Thomas, A. (2020). A critical review on the fused deposition modeling of thermoplastic polymer composites. *Composites Part B Engineering, 201*, 108336.

CHAPTER 4

Technological Approaches on Pharmaceutical 3D Printing

PRAVIN ADMANE[1], KULDEEP VINCHURKAR[2], and SHEETAL MANE[3]

[1]*Department of Pharmacy, Gurunanak Technical Institute, Nagpur, Maharashtra, India*

[2]*Department of Pharmaceutics, Sandip Institute of Pharmaceutical Sciences (SIPS), Affiliated to Savitribai Phule Pune University (SPPU), Pune, Nashik, Maharashtra, India*

[3]*School of Pharmacy, Devi Ahilya Vishwavidyalaya, Indore, Madhya Pradesh, India; NMT Gujarati College of Pharmacy, Indore, Madhya Pradesh, India*

*Corresponding author

ABSTRACT

In current scenario, pharmaceutical industries are more focused on manufacturing of personalized medicines using various computer-assisted technologies. Some of the dosage forms manufactured by traditional method are now being changed to a new concept of formulation technologies, which enable the manufacturing of dosage forms as per consumer demand and satisfaction. 3D printing of pharmaceuticals is such a novel technological approach that is used to produce 3D-printed dosage forms by depositing printing ink comprising the drug using computer-assisted design. As compared to traditional manufacturing of dosage forms, 3D printing methods allow the manufacturing of dosage forms comprising more complex drug combinations. It also allows tailoring of personalized dosage forms according to patient requirements, having flexibility in shape and design according to market demand, which improves patient

Advancements in Fused Deposition Modeling (FDM) Technology for Pharmaceutical and Medical Applications. Kuldeep Vinchurkar, Dinesh Kumar Mishra, Pankaj V Dixit, and Sheetal Mane (Eds.)

DOI: 10.1201/9781779640338-4

compliance with dosage forms. As the formulation of 3D-printed drug delivery systems offers personalized dose, it may offer several advantages in different areas of the healthcare system, including hospitals, community pharmacies, the PHC system, etc. 3D-printed technologies also offer an accurate and precise dose of potent drugs, minimizing side effects and wastage of product. Using different methods of 3D printing technology, like binder jet printing, fused diffusion modeling, semisolid extrusion, direct powder extrusion, stereolithography, selective laser sintering, and various personalized dosage forms of different drugs can be fabricated in the future, considering regulatory aspects and acceptability.

4.1 INTRODUCTION

3D printing is a novel transformational technology, where a computer-aided design software is used to print polymeric ink comprising drug molecules onto a substrate to develop a product, achieving personalized dosing of medication with accuracy and precision. Three-dimensional printing works on the principle of layer-by-layer deposition of ink onto the substrate using computer design software. The technique was introduced first in the 1990s. This technology originated and was developed more than 30 years ago by the efforts of combined research from various departments including optics, robotics, and chemistry in order to construct various models using UV-cured resins as outcome product [1, 2]. The first 3D-printed tablet of levetiracetam under the brand name Spritam was approved by the FDA, after which the revolutionary exploration of 3D printing technology is being done in the field of drug delivery [3, 4]. The working of 3D printing technology involves the ejection of drug ink from the printer head onto X–Y plane to create the base of the substrate. Then the printer ejects a liquid binder onto the base by traveling lengthwise through the Z-axis. The process is being performed with the help of computer software and repeated till the desired thickness is achieved, still the substrate is customized layer-by-layer to achieve the desired thickness. Using 3D printing technology, highly reproducible pharmaceutical dosage forms can be easily created by accurately controlling of droplet size of ink and obtaining complex drug release profiles in a precise way [5]. Various 3D printing technologies have emerged during research based on different principles and working mechanisms [6]. As per the American Society for Testing and Materials, 3D printing technologies are classified into seven categories on the basis of their principles, namely binder jetting, vat photopolymerization, sheet lamination, material extrusion, directed energy deposition, powder bed fusion, and material jetting [6, 7]. These technologies offer flexibility in printing design, tailoring of personalized drug doses and drug combinations, instant manufacturing on demand, dispensing low volumes accurately, enabling release of drug in a precise manner to achieve desired pharmacological action, minimizing side effects, and wastage of ingredients [8, 9]. In addition, these 3D printing technologies have been utilized to manufacture a dosage form with different drug release profiles such as controlled-release formulations, immediate-release formulations, delayed release formulations, oro-dispersible films, transdermal patches, implants, etc. [9].

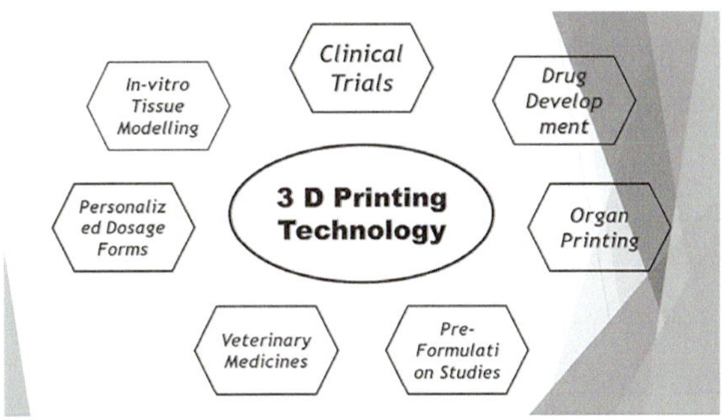

FIGURE 4.1 3D Printing technologies and their applications.

Source: Reprinted from Ref. [10]. https://creativecommons.org/licenses/by/4.0/

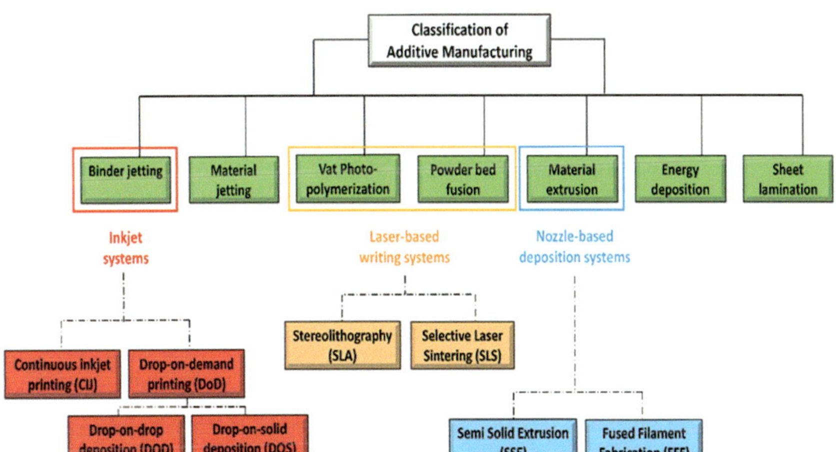

FIGURE 4.2 Classification of 3D printing techniques.

4.2 CLASSIFICATION OF 3D TECHNIQUES

The detailed classification of 3D printing techniques is given in Figure 4.2. 3D printing techniques are classified into three major categories:

1. Printing-Based Inkjet Systems
2. Laser-Based Writing Systems
3. Nozzle-Based Deposition Systems

1. **Printing-Based Inkjet System** involves deposition comprising different combinations of drug and excipients layer-by-layer into a non-powder substrate, where ink is sprayed in varying sizes of droplets [11].
2. **Laser-Based Writing System** works on the principle of photopolymerization, where the ultraviolet light beam is passed through the surface of the liquid resin to a previously designed 3D model.
3. **Nozzle-Based Deposition System** involves deposition of ink material layer-by-layer through the nozzle extruder onto the specified substrate as per computer-aided software design or pattern.

The above systems are further classified into the following subclasses:

1. Printing-based Inkjet Systems
   - I. Continuous Inkjet Printing (CIJ)
   - II. Drop on Demand Printing (DoD)
     - a. Drop on Drop Deposition (DoD)
     - b. Drop on Solid Deposition (DoS)
2. Laser-based Writing Systems
   - I. Stereolithography (SLA)
   - II. Selective Laser Sintering (SLS)
3. Nozzle-based Deposition Systems
   - I. Semi-Solid Extrusion (SSE)
   - II. Fused Filament Fabrication (FFF)

4.2.1 PRINTING-BASED INKJET SYSTEMS

4.2.1.1 CONTINUOUS INKJET PRINTING (CIJ) [12]

CIJ printing enables deposition of ink with a drop diameter approximate up to 100 μm. This technique is applied for the purpose of coding and marking.

In the CIJ printing technique, the stream of liquid drops is formed as per the Rayleigh instability theory of a liquid column. The liquid drops are being forced out through a small nozzle under pressure. The nozzle is set at a specific position relative to the horizontal surface, so as to develop a minute charge on each drop as formed, which is required to deposit the drops at a specific position. Further, individual drops of liquid are guided by generating potential on deflector plates. When multiple nozzles are being used, a minute fluctuation of pressure is generally applied behind the nozzle with the help of a piezoelectric transducer. This process is beneficial for synchronization of liquid drops to be deposited, as multiple nozzles are working in parallel. Figure 4.2 illustrates a diagrammatic representation of a CIJ printer. The unwanted drops of ink which is produced are then guided to a gutter with the help of an electric field, which may be recycled in many applications like graphics, product marking, etc. The recycling process may sometimes cause contamination of the ink due to exposure to environment. The deposition of drop at specific position is controlled by either guiding a drop away from substrate or by positioning it onto the substrate. This process is called binary printing, in which the drop of ink is positioned either in the recycling gutter or onto the substrate. These CIJ systems generate drops in the region of 20–60 kHz with as drop velocity of more than 10 ms^{-1} at nozzle.

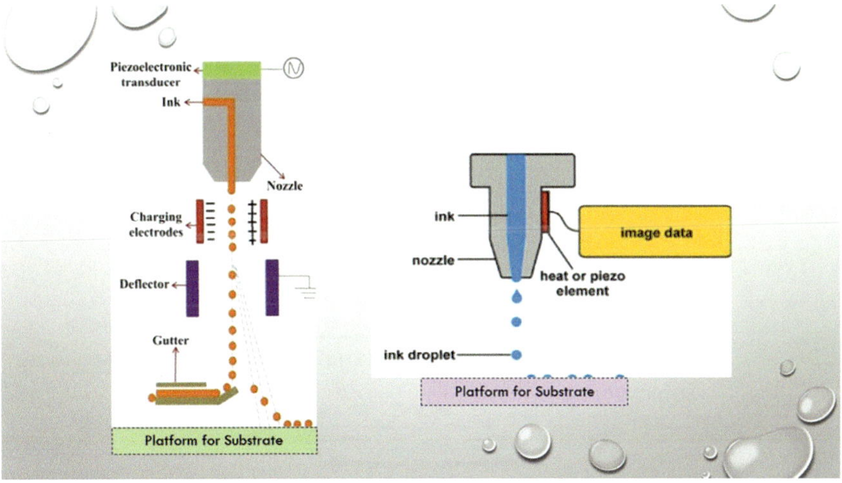

FIGURE 4.3 Schematic diagram of: (a) Continuous inkjet printer and (b) Drop-on-demand printing.

Source: Reprinted from Ref. [13]. https://creativecommons.org/licenses/by/4.0/

4.2.1.1.1 Drop-on-Drop Deposition

In this technique, the head of the printer ejects the droplets of the ink, which causes the deposition of drops onto each other. This will produce a solid layer and form high resolution 3D structure. This printing technique has the potential to formulate microscopic drug delivery systems with different geometries of 3D structure. The droplet size produced by this technique has a diameter of approximately 100 μm. The thickness of the solid layer is smaller than the size of the droplet, which may be due to solvent evaporation, shrinkage, or surface wetting [14]. The printable ink should be suitable for jetting purposes, and it should be solidified immediately. The printable ink should acquire optimized physical properties such as viscosity and volatility, which are important factors to prevent the fluid from leaking, clogging of the nozzle, and the coffee ring effect. The optimum viscosity of printable ink reported is in between 8 and 14 centipoise [15]. Some of the factors, including physicochemical properties of the drugs, need to be considered as these may affect the drug loading capacity and stability of the product, thereby affecting the therapeutic efficacy of the product [16].

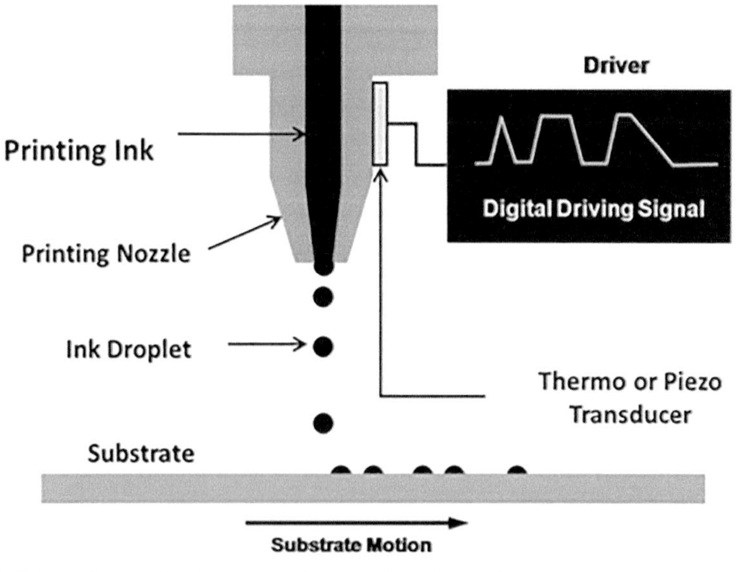

FIGURE 4.4 Schematic diagram of drop-on-drop deposition.

Source: Reprinted from Ref. [17]. https://creativecommons.org/licenses/by/4.0/

4.2.1.1.2 Drop-on-Powder Deposition

Drop-on-powder deposition is also known as drop-on-solid or drop-on-bed deposition, plaster printing, binder jetting, or powder bed 3DP, as the binder (liquid ink) is being sprayed onto the bed of a powder layer kept on the platform. As compared to drop-on-drop deposition, drop-on-powder deposition is more utilized for printing of drugs in a wide range, that is, from chemical molecules to biomolecules [16, 18]. Further, the fresh powder layer is again spread onto the platform by lowering the platform, and again the liquid ink is sprayed. The process is repeated until a 3D structure is generated [16]. The powder bed is kept at a height of 200 μm with a particle size ranging from 50 to 100 μm. The droplets adhere to each other on solidification of the binder ink, which ultimately forms a 3D object [18]. The factors, like layer thickness and spacing, should be optimized to improve adhesion between the two layers. This technology was initially developed at an institute named MIT (Massachusetts Institute of Technology), and it was patented by Emanuel Sachs in 1993. He developed the technique by depositing glycerin/water binder onto a gypsum-type powder bed using thermal bubble inkjet printheads [19, 20].

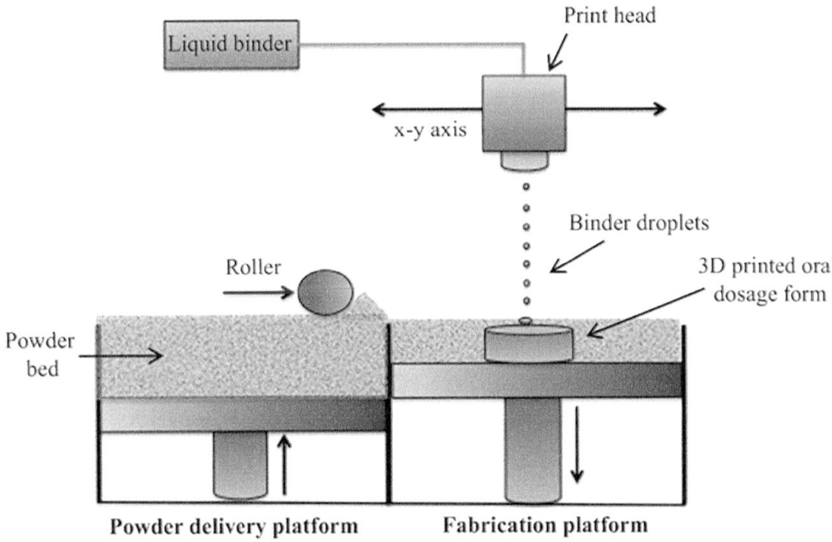

FIGURE 4.5 Drop on solid deposition 3D printing.

Source: Reprinted from Ref. [5]. Copyright © 2023 by the authors. Licensee MDPI, Basel, Switzerland. https://creativecommons.org/licenses/by/4.0/.

4.2.2 LASER-BASED WRITING SYSTEMS

This system works on the principle of photopolymerization, in which successful interaction between the ultraviolet light and photo initiator causes the release of free radicals, which are responsible for the formation of a 3D object. The system comprises the following types.

4.2.2.1 STEREOLITHOGRAPHY (SLA)

SLA is based on the laser printing method, which acts as a solid free form fabrication method. In this technique, a laser consisting of a high-energy ultraviolet light beam is passed over the uppermost part of a liquid resin, which makes controlled solidification of resin by the process of photopolymerization [21]. First, a movable platform is filled with liquid photopolymer and is kept in the vessel, which is then lifted near the liquid photopolymer surface. Then the appropriate laser is applied onto the platform so as to solidify the liquid material. Once solidification is done, the platform is then moved downwards into a vessel beneath the new polymerized layer, equivalent to its thickness. The process is continued till a solidified 3D structure is formed [22]. Due to high resolution, SLA is able to create a complex structure, and also requires low heating during the process. Thereby, this technique is highly useful for thermo-labile drugs. The 3D object prepared by SLA offers good surface quality. The selection of liquid photopolymer should always be based on its solidification after exposure to ultraviolet rays. Also, such a liquid polymer should be approved by the Food and Drug Administration for human use. But, due to a lack of approval by the FDA and low drug loading capacity, it limits the use of SLA in the pharmaceutical sector [23]. This technique has already been applied to formulate oral dosage forms comprising 5% paracetamol and 2.5% aspirin. It is also widely used for implants and in tissue engineering [23, 24]. Research studies revealed that this technique may be utilized for the manufacturing of multi-layered tablets comprising multiple drug molecules to achieve personalized medication as per patient need [24, 25].

4.2.2.2 SELECTIVE LASER SINTERING

SLS involves the melting of the powder raw materials using a high-power laser, followed by the fusion of the melted powder materials together [26]. SLS uses an energy source in the form of high-power laser to make a

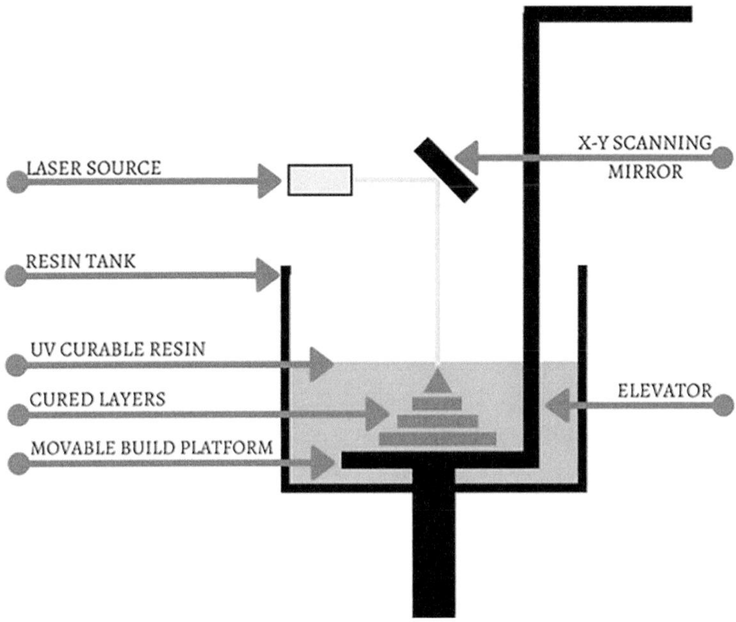

FIGURE 4.6 Diagrammatic representation of the working of SLA.

coherent mass of powder photopolymer kept on a platform. After fusion of powder photopolymer, the platform moves downward so as to refill the platform with powder, and then the process is being further continued. SLS 3D printing has been widely used by engineers and manufacturers for decades. Due to its low cost per part, established materials, and high productivity make this technology is ideal for use in various applications from rapid prototyping to small-batch, pilot batch, or commercial manufacturing. SLS is working based on the following steps.

1. **Printing:** The powder material is first spread on the top of a platform to form a thin layer inside the build chamber. The powder is preheated by the printer to a temperature somewhat below the melting point of the raw material, so that the laser can easily raise the temperature of the selected portion of the powder bed. The laser first scans the 3D pattern of the object, and then heats the powder to just below or right at the melting point of the material. This makes the fusion of all the particles together mechanically to form a single coherent mass, and forms one solid part. The unfused or unsintered powder forms support structures by surrounding

the parts during the further printing process. So, separate support structures are not required in this process. The platform is then moved downwards approximately between 50 and 200 μm by one layer into the build chamber, and this process is repeated for each layer till the completion of the cycle.

2. **Cooling:** After completion of the printing step, the build chamber should be slightly cooled down in two steps, first the inner part of the print enclosure and later on the outside of the printer. This ensures the building of optimal mechanical properties and also avoids warping in any part of the object.
3. **Post-processing:** The finished parts will then be removed from the build chamber. It will be separated, and excess powder will be cleaned. The excess powder can be recycled for further repeated processes. Further, the printed parts can be post-processed using some processes of media blasting or media tumbling.

SLS 3D printing systems used in industries utilize a single or multiple high-power lasers. Generally, for the printing process, an inert environment is required, which can be maintained using nitrogen or other gases, so as to prevent powder from oxidizing and degrading. The specialized air handling equipment is required to acquire the HVAC system and industrial power. These machines, even the smallest industrial machines, require at least 10 m^2 of installation space.

SLS printing technique allows availability of such a platform, which makes easy printing of overhanging features, interlocking parts, intricate geometries, along with some other complex designs. SLS has the potential to make new design and engineering, due to its viable rapid manufacturing method. SLS 3D printers are capable of producing complex geometries very efficiently that are impossible or costly to manufacture with traditional processes. SLS also has the ability, in case of complex assemblies, which normally require multiple parts to be operated into a single part. SLS can operate in its full potential due to its lightweight designs, so that complex lattice structures can be easily manufactured, which is impossible to manufacture with traditional methods.

SLS offers several advantages, like it has more chemical resistance, can prepare products of higher strength and provide faster printing speed. SLS technique has been used to develop various types of drug delivery systems including oral dosage forms with controlled release profile, miniprintlets of single and combinational drugs like paracetamol and ibuprofen [27].

SLS is also utilized for a wide range of materials, including ceramics, polymers, metals, etc.

4.2.3 NOZZLE-BASED DEPOSITION SYSTEMS

Nozzle-based printing systems are being used to overcome limitations of inkjet printing systems including printing only low therapeutic doses, difficulty in multiple layer printing, insufficient hardness, formation of a rough surface, and longer drying time [28, 29]. In this system, the binder solution is first prepared in which the components comprising drugs, suitable polymers, along with remaining solid ingredients are uniformly mixed. This mixture is then sprayed through a nozzle and deposited to form the subsequent layers directly to create a three-dimensional object [30]. This technology is classified into two basic methods, one is fused deposition modeling (FDM) and another is pressure-assisted microsyringes (PAM).

4.2.3.1 FUSED FILAMENT FABRICATION (FFF)

FFF is one of the most frequently used 3D printing system, also known as fused diffusion modeling (FDM). In this system, first the drug is properly mixed in thermoplastic polymers, then kept for incubation in a suitable solvent. Or both drugs and thermoplastic polymers are melted together with the help of hot-melt extrusion technique at required temperature to form a molten mass, which is further extruded into filament using a high-temperature nozzle. The layers formed after extrusion are immediately deposited onto a plate and solidified instantly [29]. The current technique is a low-cost manufacturing process and offers several benefits as it is capable of manufacturing products using highly complex drugs with difficult structures and geometries and also provides good mechanical strength. It also helps to produce 3D-printed objects with modified release drug profile [30]. FDM printer has the potential to develop gastroretentive floating 3D-printed tablets using hot-melt extrusion in which the drug-loaded filaments are created first. This extrusion-based technology has also been used to design diversified drug delivery systems to get the desired release profile of the drug. Along with these several advantages in pharmaceutical applications, some drawbacks exist with this process like lack of sufficient quantity of biodegradable thermoplastic polymers acquiring ideal characteristics of viscosity and melting parameters required for extrusion process, another is that high temperature in extrusion process is not suitable for thermosensitive drugs.

The working of FDM 3D printing is shown in Figure 4.7. As per design protocol, 3D printer controls its nozzle position, which then spurts out melted thermoplastic substance or solidified metal powder in solid form onto a plane sheet of working platform, which will be solidified instantly to form thin layer. After solidification of the first thin layer, the nozzle of 3D printer gets back to its previous position as per design to form another thin layer outside the first layer. The above process is repeated till the completion of the process and to get the 3D-printed structure [31]. The FDM 3D printer is a rapid prototyping printer mainly based on printing plastic exhibiting plasticity. To control the structure of a 3D-printed object, FDM 3D printer is divided into two layers: first is the host computer and the second is the bottom control. The host computer is basically utilized to operate three-dimensional designing software, printing control software, slicing software, etc. The bottom control includes an embedded main board, a microcontroller, stepping motor, a limit switch, motor driver, printing platform, a thermoplastic material extruder, a temperature sensor, etc. The main and important working segments of the FDM 3D printer are design of printing nozzle and its mechanism, working platform, the heating work mechanism, the motion mechanism, and wire feeder.

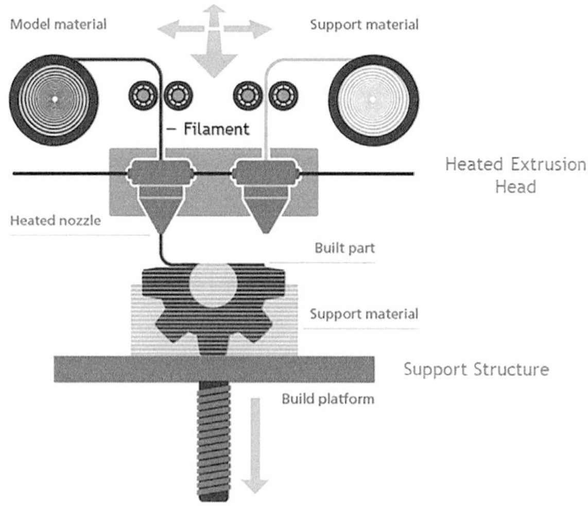

FIGURE 4.7 Schematic diagram of FDM process.

Source: Reprinted from Ref. [32]. © 2025 by the authors. Licensee MDPI, Basel, Switzerland. https://creativecommons.org/licenses/by/4.0/.

A. Nozzle of FDM 3D printer

In the FDM process of 3D printing technology, after melting the filamentous thermoplastic in nozzle's heating chamber, non-melted piston forces the melted material into the nozzle cavity under a certain pressure. The nozzle then squeezes out the thermoplastic filament of a specified diameter so as to form a highly precise finished product. The intelligent nozzle mainly includes heating pipe, fin, feeding pipe, nozzle, exhaust fan, and thermistor. The working of FDM printer involves the following mechanism, in which the printing raw material of the FDM printer first passes into the feeding pipe with the help of a wire feeder. Later on, printing raw material across the transition pipe is passed into the heating pipe. The heating temperature in the heating pipe is regulated using feedback regulation that heats the thermoplastic material via a thermistor. The heating material and melting parts then squeeze out the thermoplastic filament from the nozzle having a predetermined diameter. The nozzle structure is influenced based on its internal flow structure, the regulation of heating temperature and also the variation in pressure difference generated in different flows. Furthermore, FDM printing is also affected by change in various factors because of the regulation of temperature of nozzle. The factors include accumulation resistance, cohesiveness, extruded wire width, wire flow, etc. The low temperature leads to blockage of nozzle due to partial solidification of the material, resulting in too large viscosity of material, which ultimately will affect the velocity of wire. If the temperature is too high, it produces a partially liquid material, which results in an increase in mobility and fast extrusion speed, which ultimately leads to low precision of the formed filament. The nozzle is basically made of Teflon material, which is almost insoluble in all solvents and has acid–base and various organic solvent resistance characteristics [33].

B. The feeding mechanism of the wire of FDM printer

This process includes the wire feeding mechanism, which carries the raw material toward the nozzle of the printer. The raw materials for the FDM printing used are generally polylactic acid (PLA) plastic, nylon glass fiber, ABS plastic. The raw material diameter is generally about 1–2 mm, while the diameter of nozzle is about 0.2–0.5 mm. This diameter difference requires certain pressure in a nozzle so as to extrude the molten material at a specified speed. The principle of the wire feeder is developed based on two DC motors controls, first is the wire feed speed, and second is the open–close condition with the D/A control module, which needs to be coordinated at any moment. Wire feeding mechanism should be facilitated with push–pull mechanism so as to maintain the stability and reliability of feeding wire and to avoid the broken wire [34]. The wire feeding speed is maintained from 0 to 300 s^{-1} by applying 1/16 stepping motor of 1.8 degrees. Driving force of driving wheels and pressure of non-melted wire materials affect the formation of filamentous structure. The driving force of wheels is mainly improved by increasing the power of the stepping motor, modification in the junction type of driving wheels, and extending the friction time between driving wheels and wire materials.

C. The principle of the moving mechanism of the FDM printer

FDM printer works by travelling along X and Y axes in a planar motion and through Z-axis in a vertical motion. The principle of 3D printing is based on the conversion of an arbitrary complex three-dimensional model into plane plane-structured accumulation. The moving mechanism is mainly categorized into two types, that is, frame-type and the delta-type. The main aim of 3D printer is to convert three-dimensional printing to two-dimensional printing, hence, there is no need for three or more axes linkage. The X–Y axes linkage is used for plane printing, while the Z-axis moves the working platform at a specific height to direct the feed deposition layer-by-layer till completion as per software design. It involves the three stepping motors. The stepping motor helps to move the synchronous belt on the slider with the help of the belt wheel fixed on the motor's light bars. This makes the conversion of rotary motion of the motor shaft into rectilinear motion. The synchronous belt is responsible for moving the slider up and down on the guide. The delta-type 3D printer saves the printing space by modifying the printing height and transforming rectangular space into Cartesian space. The delta-type 3D printer differs from other printers, in which stepping motors are fixed on triangular frame required for a wire feeding mechanism, which do not move along with the nozzle simultaneously. This decreases the inertia of the nozzle, and leads to improved printing speed and its precise controlling capability.

D. The mechanism of temperature control of FDM Printer

As FDM printing process is based on heating and melting of raw material and its subsequent layer-by-layer printing through nozzle in a plane, it is necessary to control the heating temperature of nozzle and modeling temperature of the working platform. The heating rod is responsible for heating the chamber of the nozzle at predetermined temperature regulated by temperature controller through the thermistor, which controls the heating circuit in a timely feedback. The heating as well as feedback speed is mainly controlled by regulating the nozzle temperature. The raw materials used are mainly PLA plastic and ABS plastic, whose melting point is about 170°. By improving the heating speed and temperature feedback ability, the printing time can be easily shortened with improvement in the printing precision. The temperature of working platform should be controlled properly to avoid the warp and molding defects during the formation of the product. So, the working platform is stabilized at a relatively stable temperature by controlling the thermistor within specific temperature range.

The mechanism of the FDM printing technique is given in Figure 4.8.

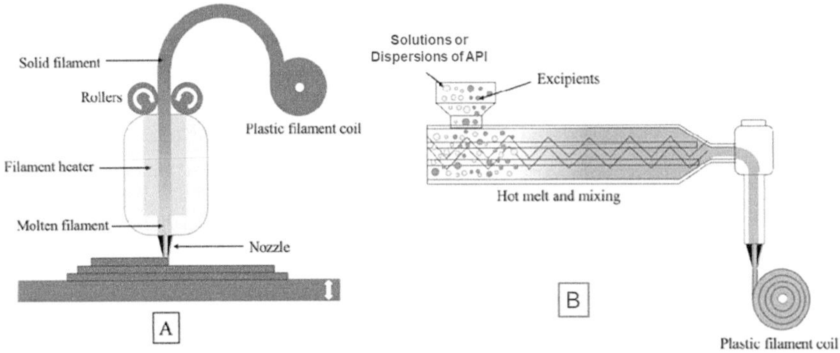

FIGURE 4.8 Schematic diagram of FDM technology in tablet preparation, (A) Printing principle of FDM technology and (B) Preparation of drug filaments using HME-FDM method.

4.2.3.2 PRESSURE-ASSISTED MICROSYRINGE (PAM)

Pressure-assisted micro syringe technique is another nozzle-based 3D printing technique used to deposit viscous and semiliquid materials onto the substrate [15, 35]. The pressurized air piston is being used to extrude the ink material from the syringe extruder. The ink material will further deposit onto the substrate layer-by-layer in a predetermined pattern in computer software. The droplets produced by PAM method are in the range of 5–10 μm or less [36]. The droplet size is dependent on the viscoelasticity parameter. This technique has the potential to develop drug delivery systems comprising a complex structure. The technique is advantageous over other techniques as it is able to extrude ink material in a continuous manner at room temperature. So, thermosensitive products can easily be fabricated with this technique. But, the solvents being used in the preparation of ink may cause safety and stability issues during production and drying steps [15, 29]. PAM technique can be utilized in a wide range of applications, including tissue printing and soft tissue scaffolds [37].

In the attempt to control the drug release, chitosan is used as a polymer in order to develop polylactic-coglycolic acid / nano-hydroxyapatite (PLGA / nHA) scaffolds. Sustained release formulation of recombinant human bone morphogenetic protein 2 (rhBMP-2) is prepared using nano-structured carriers comprising tissue-engineered bone, which is used to treat defects in large jaw. The chitosan is obtained from crustaceans shells of shrimps and crabs. The polysaccharide consists of glucosamine and N-acetylglucosamine as copolymers, which is obtained by the partial deacetylation of the chitin. This cationic polyamine shows chemical reaction with anionic systems. As chitosan is easily biodegradable and biocompatible, it acts as a non-irritant and nontoxic agent. Furthethe rmore, it is having wide application to prepare various novel dosage forms including films, gels, microspheres, liposomes, tablets with controlled release profile, like colonic drug delivery, mucoadhesive drug delivery, etc.

A. Optimization of material for extrusion-based 3D printing

1. Choice of material

The polymers to be selected should acquire the ability to form a semisolid consistency in a suitable solvent and also be able to extrude layer-by-layer from a microsyringe under the influence of pressure to create 3D structure. Generally, HPMC, PVP, ethyl cellulose, etc., are the polymers that can be used.

2. **Choice of solvent**

   Solvents play a critical role in the formation of 3D printing. Low boiling point solvents may block the tip of the extrusion nozzle due to fast drying of the materials, while high boiling point solvents cause slow drying and lead to slow solidification of the materials, thereby the 3D structure may collapse. Solvents like methanol, IPA, acetone, ethanol, and DMSO can be used for 3D printing process.

3. **Preparation of the paste**

   The required excipients are being uniformly mixed with hydrophobic polymer by the addition of a required solvent up to the optimized quantity, so as to form a homogeneous paste which can be easily extruded. The paste should be smooth and should not contain any lumps or large particles. This will prevent the blockage of the tip of the nozzle and ensure uniform flow through the nozzle. The materials to be extruded are being optimized for proper drying rate so as to print drug preparation as per predetermined design layer-by-layer, acquiring good strength.

B. Optimization of process parameters for extrusion-based 3D printing

1. **Printing pressure and printing speed**

   The printing pressure should be optimized to get the desired droplets of the ink, which is based on the extrusion ability of the drug material. The printing speed of the instrument is nothing but the expulsion of the extruder in correspondence with time (distance/time). This is expressed in the unit as mm/sec. The printing speed is optimized based on the consistency of the material and pressure.

2. **Determination of material extrusion rate**

   The extrusion rate of the material is nothing but the amount of material being extruded through the nozzle. It is based on printing speed, which depends on the pressure applied through the nozzle, which is expressed in mg/min.

3. Nozzle size and shape of the extruder

The nozzle size is to be selected based on the viscosity of the preparation and design pattern. The dimensions of different sizes of nozzles and their specifications are given in Table 4.1.

TABLE 4.1 Dimensions of Printing Nozzles.

Nozzle Gauze	Inside Diameter in μm	Length in Inches
18	840	1.25
20	630	1.25
25	420	1.25
22	280	1.25
27	203	1.25

C. Optimization of printed design for extrusion-based 3D printing

The printing design geometry was developed by computer-aided design software, and the analysis of the printed layer is being done. The STL files of the design are used to form a G-code using a Repetier-Host. These G-codes are being used as a command or feed for 3D printing of the product. The different substrates can be used for the printing of the drug material in a 3D pattern. These substrates may be made up of glass, polymers, papers, etc. The selection of substrate plays a very crucial role in depositing and adhering ink for a long period of time. Once the 3D printing process is done, each layer is evaluated in both wet as well as dry state to optimize the processing parameters. This evaluation is beneficial for the determination of total printing time. It is helpful to determine the number of layers required to obtain a complete printed structure and to determine the total printing time. The printed layers are then evaluated to determine the dimensions of the product.

4.3 POLYMER SELECTION FOR 3D PRINTING [35]

Selection of polymer plays a very important role in getting different release patterns of the drug from the 3D-printed product. The release pattern is decided depending on the patients who need to achieve the

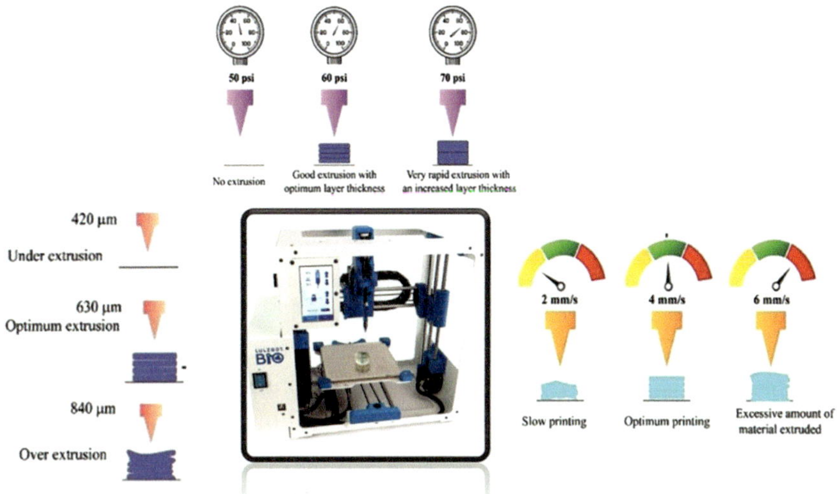

FIGURE 4.9 Illustration depicting the influence of printing pressure, printing speed, and nozzle size on extrusion-based 3D printing design for drug delivery application.

Source: Reprinted with Ref. [10]. https://creativecommons.org/licenses/by/4.0/

desired therapeutic efficacy. Different release patterns can be fabricated, comprising immediate release, delayed release, controlled release, etc. In fused diffusion modeling, two methods, named hot-melt extrusion (HME) and impregnation, are being used. Out of these, the HME technique allows higher drug loading than the others. However, the selection of polymers is quite difficult as they may be incompatible with other ingredients or processes. In the impregnation method, drug loading is done by dipping the filament in an organic solution for a specified period of time. Kollidon® VA64, PVP are good polymers that can be used for an immediate release profile. Various factors should be considered for the selection of proper polymers, including drug-polymer compatibility, drug and polymer melting point, and extrusion ability of the polymer. Polymers like polycaprolactone (PCL), polyvinyl alcohol (PVA), and PLA are also being used to develop 3D-printed products, imparting sufficient thermoplasticity. The polymer with the required grade needs to be chosen not only for the filament development but also for developing 3D-printed products, acquiring smooth and uniform extrudability properties. Polymers like HPC, Eudragit RS, RL, L, chitosan, etc., are being used to achieve modified release properties [38, 39].

4.3.1 POLYVINYL ALCOHOL (PVA)

PVA is a thermoplastic polymer. It is a synthetic water-soluble polymer having good mechanical properties and no odor. It is prepared by either full or partial hydrolysis of polyvinyl acetate, in which the acetate group is removed [40]. The melting point up to 180°C is partially hydrolyzed, whereas the melting point up to 220°C is fully hydrolyzed. The viscosity, ranging between 3.4 and 52 mPa, is partially hydrolyzed, while the viscosity from 4 to 60 mPa is fully hydrolyzed [40]. PVA has an 85°C glass transition, and the degradation temperature is about 350–450°C. This polymer is suitable due to its fewer side effects and good biodegradable property [40]. The polymer is being widely used in inkjet printing method along with glycerin to prevent blockage in the nozzle and allow smooth printing. The molecular weight of the polymer may affect the printing technique in case of inkjet printing. High molecular weight polymers are stable enough up to six months as it does not accumulate any color, whereas lower molecular weight polymer is unsuitable due to the formation of a milky appearance [41]. The polymer PVA is also showing wide application in case of FDM printing technique. Goyanes et al. successfully loaded drugs including paracetamol and caffeine along onto filaments made up of PVA using the HME method. The parameters of 200°C temperature with an extruding speed of 90 mm/s have been set for FDM printing technique.

4.3.2 POLYLACTIC ACID (PLA)

It is a biodegradable polymer used in medical fields, including regenerative medicines, drug delivery, tissue engineering, etc. The polymer is approved by USFDA for safe use in pharmaceutical products [42]. The melting point of PLA is from 150 to175°C with glass transition temperature of 55°C [42]. It is soluble in different solvents like methylene chloride, dioxane, acetonitrile, 1, 2-trichloroethane, chloroform, and 1-dichloroacetic acid. It has poor solubility in propylene glycol, water, alcohols like ethanol, methanol, and unsubstituted hydrocarbons. But solubility increases as temperature increases [42]. The PLA is not showing any carcinogenic or toxic effects. The degradation of PLA is very slow, and so, suitable for longer-term activity. It is also brittle compared to other polymers. With these advantages, PLA can be a good choice for application in 3D printing method.

4.3.3 POLYCAPROLACTONE (PCL)

It is hydrophobic in nature. It is a semi-crystalline polymer with a glass transition temperature upto 60°C having a melting point from 59 to 64°C. It is soluble in toluene, benzene, cyclohexane, chloroform, carbon tetra-chloride, etc. at room temperature. Also, it is showing low solubility in ethyl acetate, acetonitrile, dimethylformamide, and is found insoluble in alcohol, diethyl ether, etc. [43]. The degradation time of PCL is longer than that of PLA. 3D-printed polymeric nanocapsules were developed using PCl and Eudragit RL100, and then loaded with the help of an FDM printer. The extrusion temperature for Eudragit was 110°C, and for PCL was 65°C. The printing temperature of PCL and Eudragit polymer was observed at 95 and 170°C, respectively. It was found that the Eudragit tablets acquired high drug loading as compared to PCL because of its high swelling indices. The release profile in the case of Eudragit was found to be more than PCL tablets [44].

4.3.4 POLYPHENYLENE SULFIDE (PPS)

It is semi-crystalline in nature, comprising thermoplastic properties. It consists of sulfur atoms and substituted phenylene rings, which makes its asymmetrical, rigid backbone chain. It is also stable at high temperature, flame retardant, good dimensional stability, and has more chemical resis-tance property. It also has high aging resistance, more radiation resistance, and is non-toxic in nature [45, 46]. It has wider applications in the sectors of electronics, automotive engineering, etc. As the polymer has better cross-linking property, it is utilized to produce 3D printing structure with more mechanical strength [47].

4.4 APPLICATION OF 3DP TECHNOLOGY IN PHARMACEUTICAL DOSAGE FORMS

4.4.1 TABLETS

Tablet is widely used as an oral dosage formulation and can be more feasible for 3DP process. Tablets comprising a single API and multiple API have been selected for 3D printing.

4.4.1.1 SINGLE API TABLETS

Immediate release tablets comprising single API have been developed using FDM 3DP technology. A dosage form with drug loading of 60% was being developed using polymer as polyurethane which is thermoplastic in nature. The product was developed using FDM method [48]. Skowyra et al. in 2015 developed PVA filaments loaded with prednisolone using FDM method with extended release pattern with drug release up to 24 h [39]. It was found that the selection of materials and 3D printing technique affects the physicochemical properties of 3D-printed tablets and ultimately affects the release profile of the drugs. Wang et al. in 2006 developed zero-order controlled-release dosage forms of pseudoephedrine hydrochloride using 3DP technology with three different release patterns. Steriolithography and inkjet printing are the 3D printing techniques that can be used to formulate tablet dosage forms [49, 50]. For example, extended release tablets of 4-aminosalicylic acid have been successfully fabricated using the SLA technique comprising different polymers as diphenyl (2,4,6-trimethylbenzoyl) phosphine oxide, polyethylene glycol diacrylate, and polyethylene glycol at different ratios for achieving the different release profiles.

4.4.1.2 MULTIPLE API TABLETS

To achieve incorporation of complex medications with more than one API in a single tablet called a polypill, 3D printing technology can be successfully used to manufacture such a complex dosage form. First, Khaled et al. successfully developed the polypill comprising multiple APIs, including nifedipine, captopril, and glipizide using 3D extrusion-based printing. This composition is being useful to treat patients suffering from hypertension, diabetes. In this polypill, captopril was incorporated in the osmotic pump compartment, sustained release compartments of glipizide and nifedipine, and glipizide and joining layer. After administration, the quick disintegration of joining made the splitting of sustained release and captopril compartments. It was found that the zero-order drug release of captopril is due to controlled porosity in the shell based on the osmotic release pattern. The compartments of sustained release of the drugs glipizide and nifedipine showed drug release based on the principle of diffusion through the gel layers [51]. Using 3D printing methods and polymers, it is possible to create complex templates in which multiple components can be incorporated for different release patterns. Recently, Sun and Soh in

2015 developed customized tablets comprising three different components using 3DP method. The tablet prepared comprises two components of surface-eroding polymer with and without drug, respectively, and third component as an impermeable polymer, which forms a protective layer or coat. The shape of surface eroding polymer affects the drug release profile. Goyanes et al. also developed oral dosage forms of budesonide with modified release profiles using FDM method in combination with fluid bed coating and HME [30]. SLA was being used in 2016 to customize tablet dosage form loaded with both thermolabile 4-aminosalicylic acid drug and thermostable paracetamol for modified release profile. From the study, it was found that the SLA is able to reduce the degradation of the drug as compared to FDM and also allows the manufacturing of 3D printing products for thermolabile drugs [50].

4.4.2 IMPLANTS

3D-based implant is one of the novel techniques that can be utilized to achieve multi-drug loading with different drug release profiles. Huang et al. in 2007 developed an implant comprising levofloxacin drug that showed complex release patterns. This product showed a two-way profile, one with pulsatile drug release from day 5 to 25, and later steady-state release of drug from day 25 to 50, and finally again the pulsatile release pattern on day 50 to day 80. Such a technique is very beneficial in the case of bone restructuring [52]. In 2009, Wu et al. manufactured an implant using 3D printing technique consisting of multiple drugs for the treatment of bone tuberculosis. In this, drug materials, that is, isoniazid with rifampicin, were uniformly deposited in each alternate layer in a predetermined sequence to form a pattern of concentric cylinders. It was found that the drugs were sequentially released from outside to the center. This 3DP-based on multiple drugs, could be an evolutionary approach used in the treatment of tuberculosis in bone due to its dosage form and drug release pattern compatibility [53].

4.4.3 OTHERS

3DP technique is also being used for the manufacture of transdermal drug delivery systems with complex drug release profiles. Combining 3D printing technology with 3D scanning technology, it is feasible to develop more efficient transdermal patches as per the personalized demand of the patient [54]. First, the 3D scanning method is able to collect accurate and appropriate information about personal skin problems, and thereby leads to customized, tailored transdermal patches using 3DP methods. Such combinational techniques can also be utilized in the case of buccal, and so, are helpful to increase patient compliance. 3D printing technique might also be applicable in chemotherapy to treat cancer, where the drug component can be accurately targeted to the target site for diagnosis and treatment purposes. This can be done by the preparation of nanoparticles and their incorporation into a specific dosage form using a 3D printing method. However, the formation of nanoparticles using 3DP has some limits due to the unavailability of suitable material required for the preparation of nanoparticulate 3DP dosage form. As compared to traditional methods, 3D tissue models can mimic both chemical and spatial features of native tissues, they can deliver better *in vitro* and *in vivo* correlation.

4.5 CONCLUSION

The 3D printing utilized in drug formulation is a novel and evolutionary technique that plays an important role in developing drug delivery systems with different drug release profiles of complex drug combinations. This technology offers personalized drug treatment for individuals according to their needs. Recent advances in this technology also provide more safety and efficacy in drug therapy for individual patients including paediatrics, geriatrics, and patients with comorbidities. This technique offers more flexibility to incorporate multiple drugs and also offers delivery of the drugs to a specific target without alteration of drug concentration. The various 3D printing methods including inkjet printing, stereolithography, laser-based writing systems, and nozzle-based deposition systems, act as an emerging tool to develop interest for the fabrication of promising drug delivery systems benefiting the patient. In the future, these techniques can be more utilized for better drug incorporation of incompatible drugs and achieving a different drug release profile. The approach of 3D printing technology at the commercial level is beneficial not only for the fabrication of personalized doses but also for revolutionary changes in the pharmaceutical industry toward the upgradation of technologies.

KEYWORDS

- **3D printing**
- **dosage forms**
- **extruder**
- **fused diffusion modeling**
- **pharmaceutical industry**
- **solvent**

REFERENCES

1. Sachs, E. M., Cima, M. J., & Cornie, J. (1990). Three-dimensional printing: rapid tooling and prototypes directly from a CAD model. *CIRP Annals—Manufacturing Technology, 39*(1), 201–204.
2. Diogo, J. H. (2018). 3D printing of pharmaceutical drug delivery systems. *Archives of Organic and Inorganic Chemical Sciences, 1*, 2.
3. Hamburg, M. A., & Collins, F. S. (2010). The path to personalized medicine. *The New England Journal of Medicine, 363*(4), 301–304.
4. Arain, F. A., Kuniyoshi, F. H., Abdalrhim, A. D., & Miller, V. M. (2009). Sex/gender medicine. The biological basis for personalized care in cardiovascular medicine. *Circulation Journal, 73*(10), 1774–1782.
5. Shanshan, W., Xuejun, C., Xiaolu, H., Xiaoxuan, H., Xiang, L., Hui, Z., Meng, L., Zengming, W., & Aiping, Z. (2023). A review of 3D printing technology in pharmaceutics: technology and applications, now and future. *Pharmaceutics, 15*, 416. https://doi.org/10.3390/pharmaceutics15020416
6. Ishita, M., Gurvinder, K., Amir, S., Aneesah, M., & Cato, T. L. (2020). Progress in 3D bioprinting technology for tissue/organ regenerative engineering. *Biomaterials, 226*, 119536.
7. Jiménez, M., Romero, L., Domínguez, I. A., del Mar Espinosa, M., & Domínguez, M. (2019). Additive manufacturing technologies: An overview about 3D printing methods and future prospects. *Complexity, 2019*, 1–30.
8. Trenfield, S. J., Madla, C. M., Basit, A. W., & Gaisford, S. (2018). The shape of things to come: Emerging applications of 3D printing in healthcare. In *3D Printing of Pharmaceuticals; AAPS Advances in the Pharmaceutical Sciences Series* (Vol. 31, pp. 1–19). Springer International Publishing.
9. Vaz, V.M.; Kumar, L. (2021). 3D printing as a promising tool in personalized medicine. *AAPS PharmSciTech, 22*, 49.
10. Abdul, A., Mohammed, S. A., Mohammad, Z. A., & Javed, A. (2021). Optimization of semisolid extrusion (pressure-assisted microsyringe)-based 3D printing process for advanced drug delivery application. *Annals of 3D Printed Medicine, 2*, 100008. https://doi.org/10.1016/j.stlm.2021.100008

11. Sabna, K., Anroop, N., & Nimer, A. (2018). 3D printing technology in drug delivery: Recent progress and application. *Current Pharmaceutical Design, 24*, 5039–5048.
12. Brian, D. (2010). Inkjet printing of functional and structural materials: Fluid property requirements, feature stability, and resolution. *Annual Review of Materials Research, 40*, 395–414.
13. Kinga, S., Marguerite, B., Fabrice, R., Sophie, M., Steffen, Z., Jobin, V., & Manuella, C. (2022). Fabrication of modern lithium ion batteries by 3D inkjet printing: opportunities and challenges. *Heliyon, 8*, e12623. https://doi.org/10.1016/j.heliyon.2022.e12623
14. Norman, J., Madurawe, R. D., Moore, C. M., Khan, M. A., & Khairuzzaman A. (2017). A new chapter in pharmaceutical manufacturing: 3D-printed drug products. *Advanced Drug Delivery Reviews, 108*, 39–50.
15. Sumerel, J., Lewis, J., Doraiswamy, A., Deravi, L. F., Sewell, S. L., Gerdon, A. E., Wright, D. W., & Narayan, R. J. (2006). Piezoelectric ink jet processing of materials for medical and biological applications. *Biotechnology Journal: Healthcare, Nutrition, Technology, 1*, 976–987.
16. Goole, J., & Amighi, K. (2016). 3D printing in pharmaceutics: A new tool for designing customized drug delivery systems. *International Journal of Pharmaceutics, 499*, 376–394.
17. Nasim, S. (2020). Recent trends on applications of 3D printing technology on the design and manufacture of pharmaceutical oral formulation: A mini review. *Samiei Beni-Suef University Journal of Basic and Applied Sciences, 9*, 12. https://doi.org/10.1186/s43088-020-00040-4
18. Gross, B. C., Erkal, J. L., Lockwood, S. Y., Chen, C., & Spence D. M. (2014). Evaluation of 3D printing and its potential impact on biotechnology and the chemical sciences. *Analytical Chemistry, 86*, 3240–3253.
19. Amir, M., Amy, M., John, E., Fangzhou, L., Wenda, T., Corson, L., Peeyush, N., & Markus, C. (2021). Binder jet 3D printing—Process parameters, materials, properties, modeling, and challenges. *Progress in Materials Science, 119*, 100707. https://doi.org/10.1016/j.pmatsci.2020.100707
20. Sachs, E. M., Haggerty, J. S., Cima, M. J., & Williams, P. A. (1993). Three-dimensional printing techniques, US5340656 A.
21. Shankhadip, N. (2020). 3d printing of pharmaceuticals—Leading trend in pharmaceutical industry and future perspectives. *Asian Journal of Pharmaceutical and Clinical Research, 13*(12), 10–16.
22. Park, B. J., Choi, H. J., Moon, S. J., Kim, S. J., Bajracharya, R., Min, J. Y., et al. (2019). Pharmaceutical applications of 3D printing technology: Current understanding and future perspectives. *Journal of Pharmaceutical Investigation, 49*, 575–585.
23. Melchels, F. P., Feijen, J., & Grijpma, D. W. (2010). A review on stereolithography and its applications in biomedical engineering. *Biomaterials, 31*, 6121–6130.
24. Healy, A. V., Fuenmayor, E., Doran, P., Geever, L. M., Higginbotham, C. L., Y Lyons, J. G. (2019). Additive manufacturing of personalized pharmaceutical dosage forms via stereolithography. *Pharmaceutics, 11*, 645.
25. Robles-Martinez, P., Xu, X., Trenfield, S. J., Awad, A., Goyanes, A., Telford, R., et al. (2019). 3D printing of a multi-layered polypill containing six drugs using a novel stereolithographic method. *Pharmaceutics, 11*, 274.

26. Awad, A., Fina, F., Trenfield, S. J., Patel, P., Goyanes, A., Gaisford, S., et al. (2019). 3D printed pellets (miniprintlets): A novel, multi-drug, controlled release platform technology. *Pharmaceutics, 11*, 148.
27. Chia, H. N., & Wu, B. M. (2015). Recent advances in 3D printing of biomaterials. *Journal of Biological Engineering, 9*, 1–14.
28. Guide to Selective Laser Sintering (SLS) 3D Printing. (2021). Pharmalabs, pp. 1–20.
29. Goyanes, A., Chang, H., Sedough, D., Hatton, G. B., Wang, J., Buanz, A., et al. (2015). Fabrication of controlled-release budesonide tablets via desktop (FDM) 3D printing. *International Journal of Pharmaceutics, 496*, 414–420.
30. Konta, A. A., Garcia-Pina, M., & Serrano, D. R. (2017). Personalised 3D printed medicines: which techniques and polymers are more successful? *Bioengineering (Basel), 4*, 79–94.
31. Longwei Y., Huichao S., Xingtian Q., & Wei Z. (2016) The fused deposition modeling 3D printing. In *International Conference on Electrical, Mechanical and Industrial Engineering (ICEMIE 2016)* (pp. 201–203).
32. Rui, Z., Yang, X., Shiyi, L., Ping, H., Wenbin, H., Qingyuan, H., & Chunlai, S. (2016). Isotropic and anisotropic elasticity and yielding of 3D printed material. *Composites Part B, 99*, 506e513. http://dx.doi.org/10.1016/j.compositesb.2016.06.009
33. Yuqi, W., & Jun, H. (2012). The middle blank automatic transmission design of solar panel. *Technology Wind, 10*.
34. Wangrui, C., Yuxin, L., Fukun, L., Hang, Y., Heping, J. (2014). 3D printer architecture research. *Computer CD Software and Application, 2014*(16).
35. Ranjitsinh, P., & Atmaram, P. (2022). 3D printing of pharmaceuticals: approach from bench scale to commercial development. *Future Journal of Pharmaceutical Sciences, 8*, 48. https://doi.org/10.1186/s43094-022-00439-z
36. Vozzi, G., Flaim, C., Ahluwalia, A., & Bhatia S. (2003). Fabrication of PLGA scaffolds using soft lithography and microsyringe deposition. *Biomaterials, 24*, 2533–2540.
37. Byeong, J., Ho, J., Sang, J., Seong, J., Rajiv, B., Jeong, Y., & Hyo-Kyung, H. (2019). Pharmaceutical applications of 3D printing technology: Current understanding and future perspectives. *Journal of Pharmaceutical Investigation, 49*, 575–585. https://doi.org/10.1007/s40005-018-00414-y
38. Skowyra, J., Pietrzak, K., & Alhnan, M. A. (2015). Fabrication of extended-release patient-tailored prednisolone tablets via fused deposition modelling (FDM) 3D printing. *European Journal of Pharmaceutical Sciences, 68*, 11–17. https://doi.org/10.1016/j.ejps.2014.11.009
39. Melocchi, A., Parietti, F., Loreti, G., Maroni, A., Gazzaniga, A., & Zema, L. (2015). 3D printing by fused deposition modeling (FDM) of a swellable/erodible capsular device for oral pulsatile release of drugs. *Journal of Drug Delivery Science and Technology, 30*, 360–367. https://doi.org/10.1016/j.jddst.2015.07.016
40. Baker, M. I., Walsh, S. P., Schwartz, Z., & Boyan, B. D. (2012). A review of polyvinyl alcohol and its uses in cartilage and orthopedic applications. *Journal of Biomedical Materials Research Part B: Applied Biomaterials, 100*, 1451–1457. https://doi.org/10.1002/jbm.b.32694
41. Salaoru, I., Zhou, Z., Morris, P., & Gibbons, G. J. (2016). Inkjet printing of polyvinyl alcohol multilayers for additive manufacturing applications. *Journal of Applied Polymer Science*. https://doi.org/10.1002/app.43572

relentless quest for advancements. This journey, spanning from initial discovery to clinical trials and eventual commercialization, is a testament to the intricate dance between science, technology, and human ingenuity. Traditionally, pharmaceutical companies have relied on time-honored manufacturing methods to formulate and deliver life-saving medications. However, the dawn of the 21st century has witnessed the emergence of transformative technologies, reshaping the landscape of drug development and manufacturing [1].

5.1.1 THE EVOLUTION OF PHARMACEUTICAL MANUFACTURING: A HISTORICAL PERSPECTIVE

For centuries, pharmaceutical manufacturing has been steeped in tradition, relying on tried-and-tested methods rooted in craftsmanship and empirical knowledge. From the alchemists of antiquity to the modern-day laboratories of multinational corporations, the journey of a drug from conception to commercialization has been marked by a blend of scientific rigor, clinical insight, and industrial prowess. Conventional manufacturing methods, while reliable, have often faced challenges in terms of scalability, customization, and efficiency, prompting the search for novel approaches to meet the evolving needs of patients and healthcare providers [2, 3].

5.1.2 THE EMERGENCE OF ADVANCED MANUFACTURING TECHNOLOGIES

Fused deposition modeling (FDM)—a revolutionary additive manufacturing technology poised to redefine the way pharmaceuticals are formulated, fabricated, and delivered. Originating from the realms of engineering and design, FDM represents a departure from traditional subtractive methods, offering a tantalizing glimpse into a future where drugs are not merely manufactured but intricately sculpted with precision and purpose. At its core, FDM embodies the principles of additive manufacturing, where complex structures are meticulously crafted layer by layer, akin to a digital artisan sculpting a masterpiece [4–6.]

5.1.3 THE ESSENCE OF FDM: PRECISION AND INGENUITY

FDM epitomizes the marriage of innovation and ingenuity, harnessing the power of thermoplastic materials to create three-dimensional objects with unparalleled precision. The process begins with a thermoplastic filament, heated to its melting point, extruded through a nozzle onto a build platform, and cooled to form a precise layer of the desired object. This iterative process continues until the final product emerges, seamlessly integrating each layer into a cohesive whole. While FDM has long been celebrated in industries such as aerospace, automotive, and consumer goods for its ability to streamline prototyping and unlock unprecedented design possibilities, its application in pharmaceuticals heralds a new era of possibilities [5].

5.1.4 FDM IN PHARMACEUTICALS: A CATALYST FOR INNOVATION

Within the crucible of drug development, FDM transcends its role as a mere manufacturing technique and assumes the mantle of a catalyst for innovation. The integration of FDM into pharmaceutical drug development represents a convergence of disciplines—where the precision of engineering meets the complexity of biology, giving rise to a symbiotic relationship that holds the promise of transformative breakthroughs. Researchers and pharmaceutical companies alike are poised to explore a myriad of opportunities that FDM presents across the drug development continuum [7].

Foremost among these opportunities is the prospect of personalized medicine, where the one-size-fits-all approach of traditional drug formulations gives way to tailored therapies tailored to the unique genetic makeup and physiological profile of each individual patient. Through the precise control afforded by FDM, pharmaceutical manufacturers can envision a future where medications are custom-crafted to address the specific needs and preferences of patients, ushering in an era of unprecedented therapeutic efficacy and patient satisfaction [8, 9].

Moreover, FDM holds the key to unlocking the mysteries of drug solubility—a perennial challenge that has plagued pharmaceutical scientists for decades. By leveraging the flexibility and precision of FDM, researchers can explore novel formulations and delivery systems that enhance the solubility and bioavailability of drugs, thus overcoming barriers that have long hindered the efficacy of many medications [10–13].

Furthermore, the intricate architecture afforded by FDM opens new vistas in the realm of drug delivery, enabling the design of complex delivery systems capable of precise control over drug release kinetics, localization, and dosage. From implantable devices to transdermal patches, the possibilities are as vast as the imagination, offering new avenues for treating chronic conditions and improving patient compliance [14, 15].

However, amidst the promise of untold possibilities, lie a myriad of challenges that must be navigated with precision and foresight. Regulatory considerations, material selection, and scalability present formidable hurdles that must be addressed to realize the full potential of FDM in pharmaceutical drug development. Nevertheless, with collaboration, innovation, and a steadfast commitment to patient care, the integration of FDM into pharmaceutical manufacturing holds the promise of revolutionizing the field and improving the lives of millions worldwide [16, 17].

The integration of FDM into pharmaceutical drug development represents a paradigm shift in the way medications are formulated, fabricated, and delivered. With its precision, flexibility, and potential for customization, FDM offers a pathway to personalized medicine, enhanced drug solubility, and innovative drug delivery systems. While challenges abound, the opportunities for transformative breakthroughs are boundless, paving the way for a future where every patient receives the tailored care they deserve. As we stand at the precipice of this new frontier, let us embrace the possibilities and embark on a journey of innovation, collaboration, and discovery in the service of human health and well-being [18, 19].

5.2 OPPORTUNITIES OF FDM IN DRUG DEVELOPMENT

5.2.1 PERSONALIZED MEDICINE

Personalized medicine stands at the forefront of modern healthcare, aiming to tailor medical treatments to the specific needs of individual patients. FDM, with its precision and flexibility, offers a compelling avenue for realizing the vision of personalized medicine within the realm of pharmaceuticals. Here, we delve into the intricate possibilities and profound implications of utilizing FDM to fabricate customized dosage forms that are finely tuned to meet the unique requirements of each patient.

1. **Precision Tailoring of Dosage Forms:** At the heart of personalized medicine lies the ability to deliver precise doses of medication tailored to the individual characteristics of each patient. FDM empowers pharmaceutical companies to achieve this level of

precision by adjusting key parameters such as dosage strength, release kinetics, and formulation composition. Through meticulous control over these variables, drug manufacturers can create dosage forms that not only deliver the right amount of medication but also release it at the optimal rate to achieve therapeutic efficacy.

2. **Optimizing Therapeutic Outcomes:** The customization afforded by FDM enables pharmaceutical companies to optimize therapeutic outcomes for patients with varying medical needs and conditions. For instance, patients may exhibit differences in metabolism, absorption rates, or sensitivity to certain excipients present in conventional drug formulations. By tailoring dosage forms through FDM, these factors can be taken into account, ensuring that patients receive medications that are precisely calibrated to their physiological characteristics. This optimization minimizes the risk of under-dosing or over-dosing, thereby enhancing the effectiveness of treatment regimens.
3. **Minimizing Adverse Effects:** One of the key challenges in traditional drug formulations is the occurrence of adverse effects, which can range from mild discomfort to severe complications. FDM offers a solution by enabling pharmaceutical companies to design dosage forms that minimize the likelihood of adverse effects. Through careful selection of formulation ingredients and the use of controlled-release mechanisms, FDM-produced medications can be engineered to deliver therapeutic benefits while minimizing unwanted side effects. This not only improves patient comfort and compliance but also enhances overall treatment safety and tolerability.
4. **Enhancing Patient Compliance:** Patient compliance is essential for the success of any treatment regimen. However, adherence to medication schedules can be challenging, particularly for patients with complex dosing regimens or aversion to traditional dosage forms. FDM addresses this challenge by providing a platform for the creation of dosage forms that are more palatable, easier to administer, and better aligned with patient preferences. Whether it is the design of chewable tablets, flavored oral suspensions, or convenient transdermal patches, FDM enables pharmaceutical

companies to develop dosage forms that encourage patient compliance, leading to better treatment outcomes and improved long-term health.

5. **Facilitating Disease Management:** In addition to optimizing individual treatment regimens, FDM can also facilitate disease management on a broader scale. By offering flexibility in dosage form design, pharmaceutical companies can develop formulations that target specific disease states or patient populations more effectively. For example, FDM can be used to create pediatric formulations with child-friendly flavors and textures, or geriatric formulations that are easier to swallow and digest. Furthermore, FDM enables the rapid prototyping and iteration of dosage forms, allowing for agile adjustments in response to emerging treatment needs or clinical feedback.

FDM presents a transformative opportunity to revolutionize the field of personalized medicine within the pharmaceutical industry. By harnessing the precision and flexibility of FDM, pharmaceutical companies can fabricate customized dosage forms that optimize therapeutic outcomes, minimize adverse effects, enhance patient compliance, and facilitate more effective disease management. As we continue to unlock the full potential of FDM in pharmaceutical drug development, the vision of personalized medicine inches closer to realization, offering new hope and possibilities for patients worldwide [8, 20–26].

5.2.2 COMPLEX DOSAGE FORMS

FDM represents a transformative leap in the realm of pharmaceutical manufacturing, particularly in the fabrication of complex dosage forms. Traditional manufacturing techniques have often faced significant challenges in producing intricate structures with precise geometries, limiting the potential for innovation in drug formulation and delivery. However, FDM offers a compelling solution by leveraging additive manufacturing principles to overcome these limitations and unlock a myriad of possibilities in the development of complex dosage forms.

1. **Intricacies of Complex Dosage Forms:** Complex dosage forms encompass a wide range of formulations designed to address specific therapeutic needs, enhance drug stability, improve

bioavailability, and enable targeted drug delivery. These formulations may include multilayered tablets, drug-eluting implants, microparticles, nanoparticles, and other intricate structures tailored to meet the demands of modern pharmacotherapy. Achieving the desired properties and functionalities in complex dosage forms often requires precise control over factors such as drug release kinetics, spatial distribution, and physical characteristics, which can be challenging to accomplish using traditional manufacturing methods.

2. **Overcoming Limitations with FDM:** FDM offers a paradigm shift in the fabrication of complex dosage forms by enabling the precise deposition of materials layer by layer, according to digital blueprints. This additive approach allows for the creation of intricate structures with customizable geometries and precise control over composition and distribution. Unlike subtractive techniques, which rely on removing material to shape the final product, FDM builds up complex structures from the ground up, offering unparalleled flexibility and versatility in dosage form design.
3. **Multilayered Tablets:** One area where FDM excels is in the production of multilayered tablets, which are increasingly used to achieve controlled drug release profiles and combination therapies. FDM enables the sequential deposition of different drug formulations and excipients, allowing for the creation of layers with distinct release kinetics and properties. This versatility enables pharmaceutical researchers to design multilayered tablets that release drugs at different rates or target specific sites within the body, enhancing therapeutic efficacy while minimizing side effects.
4. **Drug-Eluting Implants:** Another promising application of FDM is in the fabrication of drug-eluting implants, which provide sustained release of medication over an extended period. By incorporating drugs into biocompatible polymers or matrices, FDM can produce implants with precise drug loading and release profiles tailored to the needs of individual patients. These implants offer advantages in the treatment of localized conditions, such as cancer or chronic pain, by delivering therapeutic agents directly to the site of action while minimizing systemic exposure and side effects.

5. **Enhancing Drug Stability and Bioavailability:** Furthermore, FDM can enhance drug stability and bioavailability by optimizing formulation parameters and delivery mechanisms. The ability to precisely control the composition, porosity, and surface characteristics of dosage forms allows for the creation of formulations that improve drug solubility, protect against degradation, and enhance absorption. This is particularly relevant for poorly soluble drugs or those with narrow therapeutic windows, where traditional formulations may be insufficient to achieve therapeutic concentrations *in vivo*.

6. **Targeting Specificity and Personalization:** Moreover, FDM enables the development of dosage forms with enhanced targeting specificity and personalization. By incorporating targeting ligands, nanoparticles, or surface modifications, FDM-produced formulations can selectively deliver drugs to diseased tissues or cells, reducing off-target effects and improving therapeutic outcomes. Additionally, the customizable nature of FDM allows for the fabrication of personalized dosage forms tailored to individual patient needs, optimizing treatment regimens and improving patient adherence and satisfaction.

FDM offers a revolutionary approach to the fabrication of complex dosage forms, overcoming the limitations of traditional manufacturing methods and unlocking new possibilities in drug formulation and delivery. From multilayered tablets to drug-eluting implants, FDM enables the creation of intricate structures with precise geometries, enhanced drug stability, bioavailability, and targeting specificity. As pharmaceutical researchers continue to explore the potential of FDM in drug development, the future holds immense promise for the advancement of personalized medicine and the improvement of patient care [27–31].

5.2.3 RAPID PROTOTYPING

Rapid prototyping is a critical component of the iterative process inherent in drug development. It enables researchers to swiftly test and refine various formulations and delivery systems, thereby expediting the pace of innovation and ultimately facilitating the timely delivery of new drugs to market.

FDM emerges as a game-changer in this regard, offering unparalleled advantages in terms of speed, cost-effectiveness, and flexibility in prototyping within the pharmaceutical industry [32, 33].

1. **Iterative Nature of Drug Development:** Drug development is a multifaceted and iterative process that involves numerous stages, including discovery, preclinical testing, clinical trials, and regulatory approval. At each stage, researchers must evaluate and refine drug formulations and delivery systems to optimize efficacy, safety, and compliance. The iterative nature of this process demands rapid prototyping capabilities to explore a wide range of possibilities efficiently and identify the most promising candidates for further development.
2. **Advantages of FDM in Rapid Prototyping:** FDM revolutionizes the prototyping process by enabling researchers to quickly and cost-effectively fabricate functional prototypes of drug formulations and delivery devices. Unlike traditional manufacturing methods, which may require extensive tooling, setup, and lead times, FDM allows for rapid production of prototypes directly from digital designs. This eliminates the need for specialized equipment and minimizes the time and resources required to iterate designs, accelerating the overall drug development timeline.
3. **Speed and Efficiency:** One of the primary advantages of FDM in rapid prototyping is its speed and efficiency. The additive nature of FDM enables the layer-by-layer deposition of materials, allowing researchers to create prototypes in a matter of hours or days, depending on the complexity of the design. This rapid turnaround time facilitates faster iterations and enables researchers to explore multiple design variations in a shorter period, leading to more informed decision-making and faster progress through the drug development pipeline.
4. **Cost-Effectiveness:** In addition to speed, FDM offers significant cost savings compared to traditional prototyping methods. Traditional manufacturing techniques often involve high setup costs, tooling expenses, and material waste, particularly for small-batch or custom prototypes. In contrast, FDM requires minimal setup and material waste, as it only deposits material where needed,

making it a cost-effective solution for rapid prototyping. This cost efficiency allows researchers to explore a wider range of design iterations within budget constraints, maximizing the chances of identifying optimal formulations and delivery systems.

5. **Flexibility and Customization:** Furthermore, FDM offers unparalleled flexibility and customization in prototyping. Researchers can easily modify digital designs and parameters to create prototypes with varying geometries, sizes, and material compositions. This flexibility allows for the exploration of diverse formulation strategies, such as controlled-release formulations, combination therapies, or personalized dosage forms tailored to individual patient needs. By tailoring prototypes to specific research objectives, researchers can gain valuable insights into the performance and feasibility of different drug formulations and delivery systems, ultimately driving innovation and optimization.

FDM plays a pivotal role in facilitating rapid prototyping within the pharmaceutical industry. Its speed, cost-effectiveness, and flexibility empower researchers to iterate designs quickly, explore a wide range of possibilities, and accelerate the pace of drug development. By leveraging FDM for rapid prototyping, pharmaceutical companies can streamline the innovation process, reduce time-to-market, and ultimately deliver safer, more effective drugs to patients in need [34–41].

5.2.4 IMPROVED DRUG SOLUBILITY

Poor aqueous solubility stands as a significant hurdle in drug formulation, impeding the bioavailability and efficacy of many pharmaceutical compounds. Addressing this challenge is paramount in the development of effective medications. FDM, with its unique capabilities, emerges as a promising solution to enhance drug solubility and dissolution rates, thus overcoming the limitations posed by poor aqueous solubility.

1. **Understanding the Challenge:** The solubility of a drug refers to its ability to dissolve in a solvent, typically water, to form a homogenous solution. Poor aqueous solubility is a common issue encountered in drug development, particularly with hydrophobic or poorly soluble compounds. Drugs with low solubility exhibit slow dissolution rates,

leading to insufficient drug absorption and reduced bioavailability when administered orally. This limitation can significantly impact the therapeutic efficacy of medications, necessitating innovative strategies to enhance solubility and improve drug delivery.

2. **Enhancing Solubility with FDM:** FDM offers a versatile platform for addressing the challenge of poor drug solubility through various approaches:

    a. **Incorporation of Solubility-Enhancing Excipients:** FDM allows for the precise integration of solubility-enhancing excipients into drug formulations. These excipients, such as surfactants, co-solvents, and complexing agents, can enhance drug solubility by improving wetting, dispersion, and dissolution properties. By incorporating these excipients into FDM-produced dosage forms, researchers can create formulations that promote rapid and efficient drug dissolution, thereby enhancing bioavailability and therapeutic efficacy.

    b. **Creation of Amorphous Formulations:** Another strategy enabled by FDM is the creation of amorphous formulations, which exhibit higher solubility compared to their crystalline counterparts. Amorphous formulations lack the ordered molecular structure characteristic of crystalline solids, resulting in increased molecular mobility and higher dissolution rates. FDM offers precise control over processing parameters, such as temperature and cooling rate, allowing researchers to fabricate amorphous formulations with tailored properties and improved solubility. These formulations can be particularly beneficial for drugs with inherently low solubility, offering a means to enhance dissolution and bioavailability.

Advantages of FDM in Solubility Enhancement: FDM offers several advantages for enhancing drug solubility compared to traditional manufacturing methods:

1. **Precision and Control:** FDM enables precise control over the composition, structure, and morphology of dosage forms, allowing researchers to tailor formulations to specific solubility requirements. This precision ensures uniform distribution of

solubility-enhancing excipients and facilitates the creation of optimized formulations with enhanced solubility and dissolution rates.

2. **Customization and Flexibility:** FDM allows for the customization of dosage forms to accommodate varying solubility challenges and drug properties. Researchers can adjust parameters such as excipient concentration, particle size, and formulation geometry to optimize solubility and dissolution characteristics for individual drug molecules. This flexibility enables the development of tailored solutions to address specific solubility issues, enhancing the overall effectiveness of drug delivery systems.
3. **Rapid Prototyping and Iterative Design:** The iterative nature of drug development necessitates rapid prototyping and iterative design to explore and refine solubility enhancement strategies. FDM facilitates rapid prototyping by enabling the quick and cost-effective fabrication of dosage forms with varying formulations and compositions. This allows researchers to efficiently evaluate different solubility enhancement approaches, identify optimal formulations, and accelerate the development process.

FDM offers a powerful tool for enhancing drug solubility and dissolution rates in pharmaceutical formulations. By enabling the incorporation of solubility-enhancing excipients and the creation of amorphous formulations, FDM addresses the challenge of poor aqueous solubility, thereby improving drug bioavailability and therapeutic efficacy. With its precision, flexibility, and rapid prototyping capabilities, FDM holds promise as a valuable technology for overcoming solubility-related hurdles in drug development, ultimately benefiting patients by ensuring the delivery of safe and effective medications [42–46].

5.2.5 PATIENT COMPLIANCE

Patient compliance, or adherence to medication regimens, is a crucial factor in ensuring the effectiveness of treatments and ultimately improving patient outcomes. However, maintaining consistent adherence can be challenging for many patients, often due to factors such as pill burden, complex dosing schedules, and discomfort associated with traditional dosage forms. FDM presents a promising avenue for enhancing patient compliance by enabling the design and fabrication of patient-friendly dosage forms that prioritize convenience, ease of administration, and personalized delivery [47].

1. **Understanding the Challenge of Patient Compliance:** Poor adherence to medication regimens can have significant consequences, including reduced therapeutic efficacy, disease progression, and increased healthcare costs. Factors influencing patient compliance are multifaceted and may include forgetfulness, inconvenience, adverse side effects, and difficulty swallowing pills. Addressing these challenges requires innovative approaches to drug delivery that prioritize patient comfort, convenience, and ease of use.

2. **Opportunities with FDM in Designing Patient-Friendly Dosage Forms:** FDM offers a range of opportunities for designing patient-friendly dosage forms that enhance compliance and improve treatment outcomes:

    a. **Chewable Tablets:** FDM enables the fabrication of chewable tablets with customized shapes, sizes, and flavors, making them more appealing and palatable to patients, particularly children and individuals with swallowing difficulties. By incorporating taste-masking agents and pleasant flavors, FDM-produced chewable tablets can mitigate aversion to medication and encourage regular use, thereby improving adherence to treatment regimens.

    b. **Fast-Dissolving Films:** Another patient-friendly dosage form enabled by FDM is fast-dissolving films, which dissolve rapidly upon contact with saliva without the need for water. These thin, flexible films can be conveniently placed on the tongue and dissolve quickly, making them ideal for patients who have difficulty swallowing pills or who prefer discreet and portable medication options. FDM allows for precise control over film thickness, drug content, and dissolution kinetics, ensuring consistent dosing and enhanced patient compliance.

c. **Personalized Implants:** FDM also facilitates the fabrication of personalized implants tailored to individual patient needs and preferences. These implants can be designed to release medication gradually over time, reducing the need for frequent dosing and enhancing long-term adherence. By customizing implant geometry, size, and drug release kinetics, FDM enables the creation of personalized treatment regimens that align with patients' lifestyles and therapeutic requirements.

Advantages of FDM in Enhancing Patient Compliance: FDM offers several advantages for designing patient-friendly dosage forms and enhancing compliance:

1. **Customization and Personalization:** FDM allows for the customization of dosage forms to meet the unique needs and preferences of individual patients. By tailoring dosage form characteristics such as size, shape, texture, and drug release profile, FDM enables the creation of personalized medication regimens that optimize patient comfort and adherence.
2. **Ease of Administration:** Patient-friendly dosage forms produced by FDM are designed to be easy to administer, reducing barriers to adherence and promoting regular use. Chewable tablets and fast-dissolving films offer convenient alternatives to traditional oral medications, while personalized implants provide sustained release of medication with minimal intervention, enhancing patient convenience and treatment adherence.
3. **Improved Tolerability:** FDM allows for the incorporation of taste-masking agents, flavorings, and other excipients to improve the tolerability of medications and reduce adverse side effects. By enhancing the palatability of dosage forms, FDM promotes acceptance and regular use, leading to improved patient compliance and treatment outcomes.

FDM holds significant promise for enhancing patient compliance through the design and fabrication of patient-friendly dosage forms. By prioritizing convenience, ease of administration, and personalized delivery, FDM enables the creation of dosage forms that mitigate barriers to adherence and promote regular use of medications.

As pharmaceutical companies continue to leverage the capabilities of FDM in drug development, the future holds exciting possibilities for improving patient compliance and ultimately advancing healthcare outcomes [48–60].

5.3 CHALLENGES OF FDM IN DRUG DEVELOPMENT

1. **Material Selection:** Choosing suitable materials for FDM-based drug formulation is challenging due to stringent requirements for biocompatibility, stability, and processability. Limited availability of pharmaceutical-grade polymers and excipients further complicates material selection.
2. **Process Optimization:** Optimizing the FDM process parameters, including temperature, speed, and layer thickness, is essential to ensure product quality and reproducibility. However, achieving optimal process conditions for pharmaceutical applications requires extensive experimentation and validation.
3. **Regulatory Compliance:** Regulatory agencies impose stringent requirements on the manufacturing processes and quality control of pharmaceutical products. Adapting FDM to meet these regulatory standards, such as Good Manufacturing Practices (GMP) and Current Good Manufacturing Practices (cGMP), presents a significant challenge for drug developers.
4. **Scale-Up and Commercialization:** Scaling up FDM-based manufacturing from laboratory-scale to commercial production poses logistical and economic challenges. Issues such as equipment scalability, production costs, and batch-to-batch consistency must be addressed to facilitate the widespread adoption of FDM in pharmaceutical manufacturing.
5. **Intellectual Property:** The integration of FDM in drug development raises questions regarding intellectual property rights, particularly concerning patented formulations and manufacturing processes. Companies must navigate legal complexities and establish strategies to protect their innovations while complying with existing patents [61, 62].

5.4 CONCLUSION

FDM holds immense promise for revolutionizing drug development by offering opportunities for personalized medicine, complex dosage forms, rapid prototyping, improved drug solubility, and enhanced patient compliance. However, the successful integration of FDM into the pharmaceutical industry requires addressing significant challenges related to material selection, process optimization, regulatory compliance, scale-up, and intellectual property rights. Despite these challenges, ongoing research and technological advancements are paving the way for the widespread adoption of FDM in drug development. Collaborative efforts between academia, industry, and regulatory agencies are essential to overcome barriers and unlock the full potential of FDM, ultimately leading to the development of safer, more effective, and patient-centric pharmaceutical products. FDM represents a transformative technology that promises to reshape the future of drug development, offering innovative solutions to longstanding challenges and driving advancements in personalized medicine and therapeutic outcomes.

KEYWORDS

- **3D printing**
- **drug development**
- **FDM**
- **ingenuity**
- **manufacturing**
- **precision**

REFERENCES

1. Bhatt, A. (2010). Evolution of clinical research: A history before and beyond James Lind. *Perspectives in Clinical Research, 1*(1), 6–10.
2. Rantanen, J., & Khinast, J. (2015). The future of pharmaceutical manufacturing sciences. *Journal of Pharmaceutical Sciences, 104*(11), 3612–3638.
3. Politis, S. N., & Rekkas, D. M. (2011). The evolution of the manufacturing science and the pharmaceutical industry. *Pharmaceutical Research, 28*, 1779–1781.

4. Masood, S. H. (2014). Advances in fused deposition modeling. *Comprehensive Materials Processing*, 69–91.
5. Acierno, D., & Patti, A. (2023). Fused deposition modelling (FDM) of thermoplastic-based filaments: Process and rheological properties—An overview. *Materials, 16*(24), 7664.
6. Franco Urquiza, E. A. (2024). Advances in additive manufacturing of polymer-fused deposition modeling on textiles: From 3D printing to innovative 4D printing—A review. *Polymers, 16*(5), 700.
7. Cailleaux, S., et al. (2021). Fused deposition modeling (FDM), the new asset for the production of tailored medicines. *Journal of Controlled Release, 330*, 821–841.
8. Iqbal, H., et al. (2024). Status of polymer fused deposition modeling (FDM)-based three-dimensional printing (3DP) in the pharmaceutical industry. *Polymers, 16*(3), 386.
9. Mathur, S., & Sutton, J. Personalized medicine could transform healthcare. *Biomedical Reports, 7*(1), 3–5.
10. Serajuddin, A. (2023). Challenges, current status and emerging strategies in the development of rapidly dissolving FDM 3D-printed tablets: An overview and commentary. *ADMET and DMPK, 11*(1), 33–55.
11. Bhalani, D. V., et al. (2022). Bioavailability enhancement techniques for poorly aqueous soluble drugs and therapeutics. *Biomedicines, 10*(9), 2055.
12. Khalid, G. M., & Billa, N. Solid dispersion formulations by FDM 3D printing—A review. *Pharmaceutics, 14*(4), 690.
13. Bácskay, I., et al. (2022). The evolution of the 3D-printed drug delivery systems: A review. *Pharmaceutics, 14*(7), 1312.
14. Hamza, D. A. M., & El-Sherbiny, I. E. (2024). 3D printing technology in drug delivery. In *Novel formulations and future trends* (pp. 513–546). Academic Press.
15. Kayalar, C., et al. (2023). Quality control and regulatory landscape of 3D-printed drug products. In *3D printing: Emerging technologies and functionality of polymeric excipients in drug product development* (pp. 57–75). Springer International Publishing.
16. Melocchi, A., et al. Quality considerations on the pharmaceutical applications of fused deposition modeling 3D printing. *International Journal of Pharmaceutics, 592*, 119901.
17. Henry, S., Vanhoorne, V., & Vervaet, C. (2023). Fused deposition modeling (FDM) of pharmaceuticals. In *Additive manufacturing in pharmaceuticals* (pp. 45–96). Springer Nature Singapore.
18. Milliken, R. L., et al. (2024). Application of 3D printing in early phase development of pharmaceutical solid dosage forms. *International Journal of Pharmaceutics*, 123902.
19. Serrano, D. R., et al. (2023). 3D printing technologies in personalized medicine, nanomedicines, and biopharmaceuticals. *Pharmaceutics, 15*(2), 313.
20. Englezos, K., et al. (2023). 3D printing for personalised medicines: Implications for policy and practice. *International Journal of Pharmaceutics, 635*, 122785.
21. Vaishnav, P., et al. (2023). Fused deposition modelling (FDM) in personalized medicine—An overview on the rise of fused deposition model. *International Journal of Pharmaceutical Sciences and Nanotechnology (IJPSN), 16*(2), 6453–6465.

58. Raheem, A. A., et al. (2021). A review on development of bio-inspired implants using 3D printing. *Biomimetics, 6*(4), 65.
59. Bassand, C., et al. (2023). 3D printed PLGA implants: APF DDM vs. FDM. *Journal of Controlled Release, 353*, 864–874.
60. Parulski, C., et al. Challenges of fused deposition modeling 3D printing in pharmaceutical applications: Where are we now? *Advanced Drug Delivery Reviews, 175*, 113810.
61. Warsi, M. H., et al. (2018). 3D printing methods for pharmaceutical manufacturing: Opportunity and challenges. *Current Pharmaceutical Design, 24*(42), 4949–4956.
62. Wallis, M., et al. 3D printing for enhanced drug delivery: Current state-of-the-art and challenges. *Drug Development and Industrial Pharmacy, 46*(9), 1385–1401.

CHAPTER 6

FDM's Role in the Production of Medical Devices

PRANAB NASKAR, PANKAJ PAUL, ANKIT TIWARI,
KHUBAIB AKHTER KHAN, and AMALESH SAMANTA*

Division of Microbiology and Pharmaceutical Biotechnology, Department of Pharmaceutical Technology, Jadavpur University, Kolkata, West Bengal, India

*Corresponding author

ABSTRACT

Rapid production of new medical devices (MD) has gained interest in the post-pandemic (COVID-19) world. For rapid and interest focused production of MD, some scientific procedures are needed to be followed. Fused deposition modeling (FDM) is one of the most remarkable scientific additive manufacturing techniques which are widely applied in the pharmaceutical field. FDM is generally characterized with a modern 3D printing technique which is used for the manufacturing of polymeric materials. It is easy to use, accurate for prototyping, cost-effective, time-efficient, and reduces material waste. FDM makes 3D structures through the layer-by-layer melt-extrusion of a polymeric filament. The creation of a 3D printed structure involves the drawing of a digital design of the model by a 3D design software which is then used by the printer for its construction until the whole model is recreated. With the flexibility of FDM, designers can easily and rapidly change designs without setting up additional equipment or tools. Using this technology, manufacturers can manufacture devices matched to a patient's anatomy or devices having very complex internal structures. A huge interest in 3D printing of MD is increasing due to these capabilities. During the current

Advancements in Fused Deposition Modeling (FDM) Technology for Pharmaceutical and Medical Applications. Kuldeep Vinchurkar, Dinesh Kumar Mishra, Pankaj V Dixit, and Sheetal Mane (Eds.)

DOI: 10.1201/9781779640338-6

pandemic (COVID-19), nasopharyngeal swabs for COVID-19 diagnosis, face masks, respiratory valves, and so forth have been fabricated using FDM. MD produced using FDM include orthopedic and cranial implants, surgical instruments, spinal implants made from titanium alloy, dental restoration, and prosthetic devices. FDM can produce polymeric composite structures with designs as well as compositions of interest in a range of materials according to the requirement. There are multiple steps to print a 3D device: (a) Device design, (b) Software workflow, (c) Material controls, (d) Printing, (e) Post-processing process validation verification, and (f) Testing. The frequency of steps for each device is generally dependent on factors like complexity of the device. Regulatory requirements are needed for the production of the devices using FDM technology. Few requirements are applied to MD before they are marketed and others are applied to post marketed MD. In 2016, the US FDA issued a draft guidance to advise manufacturers, who are producing devices through FDM technology to determine safety and effectiveness. Different countries classify MD into various categories depending on its risk of usage, safety, and efficacy. Recently, some countries unitedly formed some international organizations which are making regulations and guidelines for MD. It helps to maintain quality of MD and also helps with import and export.

6.1 INTRODUCTION

Thousands of years ago, many civilizations used tools or devices such as knives, needles, scalpels, trocars, forceps, saws, lancets, and cautery devices for various medical procedures. Most of the medical procedures involved treatment of injuries and wounds caused by arrows, knives, sabres, guns, and cannons during wars on the battlefields [1]. The historical origins of medical devices (MD) trace back to approximately 7000 BC in Baluchistan, where Neolithic practitioners utilized flint-tipped drills and bowstrings for dental procedures. Archaeological evidence and Roman medical texts further affirmed the widespread usage of diverse MD during the ancient Roman era [2].With the development of science in the 17th century, the devices became more prevalent. Doctors or small companies manufactured MD and sold them directly to the public without maintaining quality or standards or oversight for safety and effectiveness. The 1800s was a field-breaking era for the development of MD.

In the 19th century, MD like the stethoscope, the hypodermic syringe, the ophthalmoscope, the electrocardiogram, hearing aids, and the kymograph were brought to market. In the 20th century, there was an explosion of MD that included cardiac-defibrillators, hip and knee replacements, dialysis machines, insulin pumps, and so forth [3]. In the last 100 years, MD have grown exponentially. Polyvinyl chloride (PVC) is generally a material of choice that is used in the manufacturing of different types of MD such as IV bags and tubing, and silicone tubing for catheters. Polyolefins are generally used for trays and bottles. IV catheters are made with fluoropolymers. Also, at the same time, a breakthrough of new products was witnessed like vascular grafts, heart valves, and pacemakers in the cardiovascular space. Now, there are more than 500,000 types of MD as well as *in vitro* diagnostic devices (IVDs) on the market (FDA) [4]. According to the European Patent Office, MD represent the most active technological field in terms of the number of patents filed [5]. Till now, the requirements of new MD is a hot topic. To ensure more patient safety and effectiveness, a new concept of the term "Personalized Medical Devices" has been introduced. Personalized medical devices (PMD) are the special types of MD that are specifically designed and manufactured to meet the needs of an individual [6]. With advances in materials science, computational power, imaging, and manufacturing technologies, it is now possible to produce MD that are designed to meet the specific needs of a particular patient. The manufacturing processes of PMDs are also moving closer to the point-of-care (like 3D printing in hospitals), which serves as a huge advantage to patients and healthcare professionals. Though, a lot of new risks have also emerged with PMDs and point-of-care of manufacturing that previously did not exist for traditional mass-manufactured MD [7].

Fused deposition modeling (FDM) is one of the most advanced, effective, and modern medical device manufacturing processes. FDM is a thermoplastic extrusion process employed for additive manufacturing (AM), wherein objects are constructed layer-by-layer. This technology, notable for its versatility and affordability, has garnered significant interest across engineering and medical domains due to its capacity to fabricate intricate structures. FDM holds promise in various applications, spanning from prototyping to industrial manufacturing, owing to its low-cost, diverse material compatibility, portability, and user-friendly operation. In the medical realm, FDM presents a transformative potential by facilitating the realization of innovative concepts such as dual printing and concurrent drug delivery systems. Its precise control over design and fabrication enables the creation of customized drug delivery platforms, thereby enhancing therapeutic efficacy and circumventing limitations associated with traditional manufacturing methods [8]. The recent advancements of FDM technology in healthcare system emphasize its potential benefits in the production of pharmaceutical products while also addressing existing limitations [9]. Three-dimensional (3D) printing stands as a burgeoning manufacturing technology which is applicable to biomedical and pharmaceutical sectors. FDM represents an economically feasible extrusion-based 3D printing method capable of sequentially depositing materials to construct solid structures [10]. FDM is classified within the realm of AM processes, alongside methods such as stereolithography (SL) and selective laser sintering (SLS). AM involves the incremental addition of material to construct components. Conceptually, this can be likened to stacking a deck of cards, with each layer representing a sequential addition of material. FDM, as implied by its name, operates by depositing molten material layer-by-layer to form the desired object. This rapid prototyping (RP) technique finds applications in modeling, prototyping, and production scenarios. Originating in the late 1980s, FDM was pioneered by S. Scott Crump and subsequently commercialized in 1990 by Stratasys [11].

6.2 WHAT ARE MEDICAL DEVICES?

It is inadvertently mentioned as per the Food, Drug & Cosmetic Act (FD&C Act), a medical device is defined as an instrument, implant, apparatus, implement, machine, or other related articles, including a part or accessory which is mentioned in the official National Formulary, or the United States Pharmacopoeia. It is intended to be used in the diagnosis or mitigation or treatment or prevention of disease in man or other animals and is also intended to interfere with any structure or any function of the body. Without MD, common medical procedures—from bandaging to diagnosing HIV/AIDS, implanting an artificial hip, or any surgical intervention—could not have been possible. MD are used in every step of medical procedure by self, paramedical staff, clinicians, opticians, dentists, and by any health-care professionals in advanced medical facilities, for prevention, screening, and palliative care. Such health-care technologies are used for diagnosing illness, for monitoring treatments, for assisting disabled people, and for intervening and treating illnesses both acute and chronic [12].

Nowadays, with the development of medical science, a new idea, PMD has been added to the development of new MD. PMDs include any of the types of MD that are intended to a particular individual, which would either be custom-made, patient-matched, or adaptable medical device.

Custom-made medical device is a type of MD which meets the following criteria:

1. It is intended to be used for an individual (that can be a patient or healthcare professional).
2. It is manufactured in accordance with a written advice of an authorized healthcare professional along with specific design characteristics, and is given under his/her responsibility; even though the design might be developed in consultation with a manufacturer.
3. It is intended to specify the anatomic-physiological features or pathological condition of the patient.

Patient-matched medical device matches the following criteria:

1. It is similar to a patient's anatomy within a specified design developed using techniques such as measurements of the device based on anatomic references, or by using the full anatomical features from patient imaging.
2. It is specially made in a batch through a process that can be validated and reproduced.
3. It is designed and produced under the responsibility of a manufacturer even though the design may be created in accordance with an authorized healthcare professional.

Adaptable medical device meets the following criteria:

1. It is produced in bulk.
2. It is adapted, adjustable, assembled, or shaped at the point of care, in accordance with the manufacturer's validated method, suitable for an individual patient's specific anatomical and physiological features before use [6].

6.3 IMPORTANCE OF DEVELOPMENT OF NEW MEDICAL DEVICE

Common medical treatments would not be possible without medical equipment, including bandaging a sprained ankle, detecting HIV/AIDS, implanting an artificial hip, or performing any type of surgery. Medical gadgets find applications in a wide range of contexts, including the home, by laypeople, in sophisticated medical facilities, by clinicians and paramedical workers in remote clinics, by dentists and opticians, and by health-care professionals for palliative care and prevention. These medical technologies are employed in the diagnosis and treatment of acute and chronic illnesses, as well as in the monitoring of medical interventions and the assistance of disabled persons. With over 7000 generic device categories, there are an estimated 2 million distinct types of MD available on the global market today. Any tool, apparatus, machine, appliance, implant, reagent for *in vitro* usage, software, material, or other like or related item that the producer intends to be used, either alone or in combination, for medical purposes is considered a medical device [13].

The creation of novel MD has accelerated during the last 25 years, in part due to the quick advancement of science and engineering knowledge. Apart from fundamental biomedical and clinical research, bioengineering research which draws upon developments in the physical sciences, mathematics, and other engineering fields contributes significantly to the body of knowledge that enables the development of MD [14].

Government regulations have been engulfing the medical device business, which has resulted in an extended manufacturing time and high system development costs for new and upgraded medical equipment. As an ambitious substitute, FDM or 3D printing technology makes use of layer-by-layer material printing techniques to create a 3D printed object. AM, commonly referred to as FDM or 3D printing, is becoming a more affordable, effective, and customized option for the medical device industry. These devices include prostheses, dental implants, hearing aids, knee and hip implants created to order, and surgical equipment. Short product life and cost, reduced material waste, flexibility in design and production, and mobility for modification are all made possible by this technology. Numerous cutting-edge medical equipments that were previously challenging to make can now be produced using 3D printing. In addition to lowering tooling costs, 3D printing should speed up lead times and regulatory filings.

Manufacturers of MD might take advantage of the various commercial opportunities of this technology—FDM, which are listed below:

1. Using FDM to improve industrial processes, eliminate waste, and increase efficiency, resulting in lower production costs.
2. Using FDM to personalize MD to specific patient needs while remaining affordable and cost-effective.
3. Utilizing FDM technology to facilitate the creation of varied product designs and functionalities, enhancing R&D capabilities and innovation [15].

6.4 NECESSITY OF FDM TECHNOLOGY IN MEDICAL DEVICE DEVELOPMENT

MD are a major factor in our progress toward innovation. They find use in the field of prosthetics, orthodontics, and surgery, as well as for people with special needs. Thus, the development of customized equipment in the field of medical sciences is made easier by FDM or 3D printing technology [16].

6.4.1 *SURGERY*

Surgery is an extremely difficult procedure that frequently has numerous intricate side effects. By incorporating cutting-edge technologies, there was less danger both during and after surgery. The main goal is to decrease the amount of time and risk associated with the operation, while simultaneously increasing quality and compliance. Consequently, FDM 3D printing creates the groundwork for better surgical results in challenging situations [17].

When performing right hemi-colon cancer surgery, mesenteric arteries are constructed using FDM 3D printing technology. Three distinct groups were used in the randomized clinical trials: the control, 3D printing, and 3D image divisions. The length of the procedure, the amount of blood loss, and the number of lymph nodes were the most important variables; post-surgery obstacles, the healing time and patient compliance were secondary observations. The lymphatic system's node count increased and the overall operating time was significantly decreased, thanks to 3D printing technology. In addition, the prognosis following surgery was enhanced and the procedure's expense was significantly decreased. Researchers came to the conclusion that technology makes surgery more accessible and lowers the associated costs and complications [18].

A further study identified the use of 3D printing as a means of delivering excellent outcomes for patients with sarcoma. The cornerstone of treatment was challenging-to-manage surgical resection. By determining 3D anatomical structure of body, the technology made it possible to create 3D-model that helped to overcome the constraints of traditional therapy. To make the replica of the bone structure of the patient with primary bone sarcoma, the technology of electron beam melting approach was utilized. Twenty-three individuals were randomly implanted with titanium metal. Following a 2-year follow-up period, all patients were alive, and implants underwent evaluation for potential oncological, mechanical, and infectious issues. Even while the technology is pricey and requires a lengthy preoperative period, future advancements could lead to more user-friendly technologies and even robotic surgery [19].

6.4.2 OTHER APPLICATIONS

Biocompatible polymer filaments used in FDM printing frequently exhibit strong mechanical properties and structural strength, making them appropriate for a wide range of biomedical applications [20]. Consequently, FDM has made several medical advancements possible, particularly in personalized medicine, where it has allowed for the creation of individualized therapies, prostheses, insoles, and implants [21]. Additionally, FDM is perfect for creating scaffolds that can function in tissue engineering applications as an artificial extracellular matrix (ECM). FDM is a great option for tissue engineering because it provides accurate control over the shape, size, branching, geometry, interconnectivity, and orientation of the scaffold architecture. Here, biomaterials play the function of a 3D cell support that directs and regulates regeneration of local tissue and creates theoretically programmable microenvironments at cellular level that resemble those found *in vivo* [22].

Traditional scaffold fabrication processes lack the capacity to integrate interior architecture and manage porosity [23]. FDM offers more consistent control over the size and distribution of pores needed for tissue engineering applications [24]. Bone models can be created using the majority of biomaterials utilized in 3D printing [25]. FDM can be used to create lab models, biosensors, hardware, and extremely sophisticated bioengineered organoids [26]. The majority of 3D printed biomaterials can be utilized to release drugs, including growth hormones, antibiotics, and biometals [27]. Personalized medicine and the creation of innovative drug delivery systems can be easily accomplished if the production technology is easily adjustable according to the needs of the patient. Personalized tablets with altered release profiles can be 3D printed using FDM technology, which creates a carrier filament from polymers loaded with an active pharmaceutical ingredient (API) [28].

Due to its low printer cost, good print quality, and compatibility with filaments loaded with APIs, FDM is one of the most used AM processes in the development of pharmaceutical tablets [29]. The polymer filaments containing the API can be soaked in a solution containing the API, or the filaments can be created via extrusion for tablet printing [30]. Soaking the 3D-printed tablet in a solution that contains the API is an additional method. However, soaking methods require an extra drying step, which overall lengthens the process of making tablets. Additionally, the filament soaking approach enhances dosage consistency and permits the incorporation of extremely small amounts of API in the tablet [31].

In dentistry, dental applications like denture flasks, bites, mouth guards, and oral medicine administration devices are made using FDM, among other biomedical uses for the technology [32]. Dental medical modeling can now be produced because of FDM technology [33]. Therefore, in 1999, the first patient to benefit from FDM received treatment using a 3D printed medical model that represented the intricate anatomy of the patient's cleft palate. Dental crowns have also been created in dentistry using FDM technology [32].

6.5 CURRENT REGULATION TONS TO DEVELOP NEW MEDICAL DEVICE

The medical device manufacturing industry maintains a pivotal role in advancing healthcare across the world. In the new era, the research and development in MD have both been a curse and blessing for human beings. Although development of MD has expanded the ability of physicians for diagnosis and treatment of diseases, making great contributions to health and quality of life; misuse of MD and use of low-quality MD may ruin lives. Thus, regulation is needed for the manufacturing, sale, and usage of MD. The regulations (or standards, or norms) are needed to protect the user against the risks related to design, manufacturing, and packaging of MD.

6.5.1 REGULATORY REQUIREMENTS

The regulatory criteria for MD may be changed depending on the type of medical device and the place where it is being marketed. However, some common criteria may be applied to all MD, such as:

1. It must be effective and safe for use.
2. It must be manufactured in accordance with quality system regulations (QSR).
3. It is labeled informatively and accurately.
4. The manufacturer must adhere to the regulatory authority and recall it if any adverse event associated with it is reported.

6.5.2 PREMARKET APPROVAL

Premarket approval is mandatory from the regulatory authority before marketing. This is specially required for high-risk devices, like implants and life-sustaining devices. To obtain premarket approval, a premarket approval application must be submitted to the regulatory authority. The technical and clinical data of the medical device must be included in the application which demonstrates the safety, performance, and efficacy of the device.

6.5.3 POST MARKET SURVEILLANCE

After marketing, its safety, performance, and efficacy are being monitored. It is called post market surveillance. The adverse event reports and customer complaints (post market surveillance activities) must be collected. This data are subjected to identify and monitor the safety and efficacy/performance concerns with the device.

TABLE 6.1 List of Regulatory Agencies with Their Classification and Remarks.

Name of Country	Name of Regulatory Authority	Classification	Remarks
United States	Food and Drug Administration (FDA)	Class I (low risk), Class II (moderate risk), and Class III (high risk)	• **Pre-market approval (PMA):** PMA must be obtained for high-risk devices (Class III), which includes rigorous clinical testing and review. • **510(k) Clearance:** It may be sought by manufacturer for most Class II and some Class I devices, which signifies that the device is substantially equivalent to a device already available on the market. • ***De novo* request:** It designates the novel MD that are usually low risk and provide reproducible controls to manage risks. • **Humanitarian device exemption (HDE):** A marketing application is subjected to some profit and use restrictions, exempt from effectiveness requirements. • **Establishment registration:** All facilities of devices must be enrolled annually with the regulatory authority. • **Quality system regulation (QSR):** Manufacturers must maintain quality system during production to ensure that regulatory requirements are being maintained. • **Website:** https://www.fda.gov/medical-devices [4].
Canada	Health Canada (HC)	Class I (low risk), Class II (medium risk), Class III (high risk), and Class IV(implantable)	• **Conformity assessment:** Class III and Class IV devices require PMA. • **Website:**https://www.canada.ca/en/health-canada/services/drugs-health-products/medical-devices.htm [34].
European Union	European Medicines Agency (EMA)	Class I, Class IIa, Class IIb, and Class III, ascending in risk.	• **Conformity assessment:** An undertaking of conformity assessment and demonstration of meeting all regulatory assessment must be given by the manufacturer before affixing the CE marking. • **Quality management systems (QMS):** Must meet ISO 13485. • **Website:** https://www.ema.europa.eu/en/medical-device [35]

TABLE 6.1 *(Continued)*

Name of Country	Name of Regulatory Authority	Classification	Remarks
Japan	The Ministry of Health, Labour, and Welfare (MHLW) and the Pharmaceuticals and Medical Devices Agency (PMDA)	Devices are classified into Classes I to IV based on risk, with IV being the highest.	• **Pre-market certification/approval:** Either certification (by registered certification bodies) or approval (by the MHLW) is needed depending on the device class. PMDA reviews extensive technical and clinical documentation. • **Quality management system (QMS) audit:** Manufacturing system and quality control system must comply with ISO 13485 and ISO 14971. Labeling must include indications for use in Japanese. Post-market surveillance is also mandatory. • **Website:** https://www.pmda.go.jp/english/ [36].
Australia	Therapeutic Goods Administration (TGA)	MD are classified based on risk, ranging from Class I (lowest risk) to Class III (highest risk). There's also a separate category for active implantable MD.	• **Conformity assessment:** It is similar to the EU. • **Post-market monitoring:** The performance of medical device is being monitored by TGA to ensure safety and performance criteria. • **Website:** https://www.tga.gov.au/regulation/medical-devices [37].
China	National Medical Products Administration (NMPA)	Class I, Class II, and Class III based on risk. Class III devices are of the highest risk.	• **Registration and approval:** NMPA registration and approval are required for all types of MD. • **Clinical trials:** Especially for higher-risk devices, technical data and local clinical trials are needed. • **Website:** http://english.nmpa.gov.cn/ [38].
India	Central Drugs Standard Control Organization (CDSCO)	Class A (low risk), Class B (moderate risk), Class C (high risk), and Class D(implantable). Class C and Class D devices require premarket approval.	• **Device registration and licensing:** Before marketing, license from CDSCO is mandatory. • **Clinical trials:** Depending on the device classification, clinical trial data may be required. • **Import license:** An additional import license from CDSCO is needed for importer. • **Website:** http://www.cdsco.nic.in/ [39]

International Medical Device Regulators Forum (IMDRF) is an international forum of medical device which consists of the medical device regulatory authorities of Australia, Brazil, Canada, China, European Union, Japan, and the United States, as well as the World Health Organization (WHO). It was established in October 2011. Its motto is to build on the strong foundational work of the Global Harmonization Task Force on MD (GHTF) which accelerates international medical device regulatory harmonization and convergence.

IMDRF aims to maintain working link to regional organizations and other international entities that have a mutual understanding in medical device regulatory activities that are directly linked to the common goals of fostering global convergence, leveraging resources, and making safe and effective MD available worldwide [40].

6.6 REASON OF FDM USAGE IN MEDICAL DEVICE DEVELOPMENT

In the field of medical device, the demand of customized MD is increasing. Nowadays, MD are being produced using different advanced and modern techniques. FDM is one of the most advanced AM techniques which is booming due to its simplicity and lower cost [41]. FDM also offers some benefits such as precise and personalized manufacturing, reduced waste, cost-efficiency, and faster production times [21].

The global monetary expenditure for the healthcare sector is enlarging at a significant rate because of population explosion, the increasing prevalence of chronic diseases, and advancements of technology. In 2021, the expenditure of global healthcare was $9.6 trillion, of which a significant portion was allocated toward biomedical device. The annual expenditure on biomedical device development is approx. $350 billion globally. Proper economic design and production of biomedical devices can positively impact societies leading to cost savings, improved quality of healthcare system, better patient compliance, and bright future of the healthcare system. Affordable and easily accessible MD can diagnose and treat diseases rapidly, improving patient compliance and saving healthcare costs in the long run. The Internet of Medical Things (IoMT) and artificial intelligence (AI) are used in FDM printing devices. The IoMT and AI creates link among MD designing, production, and usage. IoMT can enable MD to collect and transmit real-time data, allow for remote monitoring of patient's health, and detect health problems early. Processing the data, AI algorithms provide valuable opinion on patients' conditions and enable more personalized treatment plans which are directly used in FDM technology [21].

MD play a significant role in our advancement of innovation. They are useful for individuals with specific requirements as well as in the fields of orthotics, prosthetics, and surgery. Thus, 3D printing technology creates new opportunities for the development of customized equipment in medical sciences. It might also be possible to design MD that are difficult to produce using traditional techniques. The Food and Drug Administration (FDA) authorized a few MD manufactured with 3D printing for sale in order to promote their use [42].

6.6.1 ORTHODONTICS

In the field of orthodontics, 3D medical imaging allows for more precise diagnosis and the creation of orthodontic device based on the coordinate system of the occlusal plane of the dental anatomy. The developments in medical imaging, FDM printing, and appliance and device customized orthodontics and dentistry undergo revolution in technology. Consequently, this technique allowed for the creation of dental models with various materials, improved comfort, and accelerated healing times. Redaelli et al. [43] used 3D printing to evaluate the orthodontic braces' feasibility. Modified polyethylene and terephthalate glycol were the polymers employed in the manufacturing process. Consequently, the printed model was used to evaluate polymers including polyethylene and polypropylene. The study's evaluation criteria includes cost, mechanical strength, biocompatibility of the prosthesis on human patients, and affordability. The results indicate that the suggested strategy satisfies the patient's present requirement. The structural model met all morphological requirements, and it was demonstrated that increased patient compliance increased mechanical strength. This flexible model can also demonstrate pharmacological action by incorporating prescription drugs. Thurzo et al. [44] created the orthodontic power arm model using 3D printing. The biocompatible materials were used to construct the model. The porous alloy kind of titanium was ideal for biological uses. The researchers used finite element modeling to increase the robustness of the power arm. The investigated design may then be constructed once the model was fixed to the tooth's shape. The construction demonstrated better qualities above standard approach, as evidenced by an 82% reduction in stress and a 7% improvement in tensile arm strength.

6.6.2 SURGERY

3D printing offers a basis for improvement of surgical outcomes in difficult settings. When this technology originally appeared, researchers and surgeons were intrigued by it, which led to its application in many different fields.

Chen et al. employed 3D printing to create the mesenteric vasculature following surgery for right hemi-colon cancer. The randomized clinical trials involved three separate groups: the control, 3D printing, and 3D imaging divisions. The most crucial information included the procedure's duration, blood loss, and lymph node count; secondary results included post-surgery complications, healing time, and patient compliance. Thanks to 3D printing technology, the lymphatic system's node count rose and the overall operation time was greatly reduced. In addition to these advances, the postoperative prognosis was recovered and the procedure's cost was dramatically reduced. Researchers found that while technology lowers associated costs and complications, it also enhances surgery accessibility.

An additional investigation found that people with sarcoma could benefit from using 3D printing. The cornerstone of treatment was surgical resection, albeit difficult to administer. Beyond the limitations of conventional therapy, the technology allowed for the creation of 3D modeled prostheses by expressing the body's 3D anatomical structure. The electron beam melting method was used to mimic the bone structure of the users who primary had bone sarcoma. A total of twenty-three people received titanium metal implants at random. All patients were still alive after a 2-year follow-up period, and implants were screened for possible infectious, mechanical, and oncological problems. Although the procedure is expensive and necessitates a long preoperative period, future developments may result in more user-friendly technologies and even robotic surgery.

The outcomes demonstrated improved accuracy of resection for a specific implantation in the field of pelvic surgery.

6.6.3 CUSTOMIZED IMPLANTS AND PROSTHESES

Digital STL 3D print files obtained from x-ray, MRI, or CT scans can be used to create prostheses and implants with nearly any desired geometry. Therefore, 3D printing can be used in the medical profession to create simple and complicated personalized prosthetic limbs as well as surgical implants, frequently in as little as 24 h. This approach has been used to create dental, spinal, and hip implants. Implants used to need a lengthy and involved validation process before they could be utilized in therapeutic settings [45].

6.6.4 TESTING DIFFERENT DEVICES IN SPECIFIC PATHWAYS

One such example is to stimulate different vascular patterns to assess the effectiveness of a cardiovascular system used to treat peripheral and coronary artery disease. Thanks to 3D printing, we can quickly produce prototypes of novel design concepts or improvements for existing devices [8].

6.7 FDM PROCESS IN MEDICAL DEVICE DEVELOPMENT

FDM became commercially available in the early 1990s, after Scott Crump, Stratasys' co-founder, patented the technique in 1989. The FDM technique requires a constant supply of thermoplastic filaments via a spool; layers of material are printed to create the part [49]. Typical materials: (plastic, metal, sand, wax, gypsum, etc.) are used in 3D printing. Before the printing of a prototype, a suitable printable material must be chosen, which is a challenging decision to make. Aside from the prompt development of prototype, there is a concern regarding the environmental impact of 3D printing. Given the growing importance of this topic, researchers are beginning to focus on the negative effects of carbon and waste footprints. Thus, the concept of "eco-friendly" material was created using biodegradable, recyclable, and compostable materials. From a technological and economic standpoint, the new product may be competitive and offer superior qualities than common materials [46, 47].

However, non-biodegradable materials dominate the RP materials industry, owing to their low cost. Materials are often assessed in three aspects: mechanical efficiency, procedures, and consistency; however, the accompanying graph displays the important selection criterion for selecting a printable material. Short presentation on the material qualities for the most prominent pure polymers used in FDM polylactic acid, polycarbonate, polyethyleneterephthalate, TPU (flexible), ABS, and nylon6) in recent years [48].

FDM is the most widely used method. In FDM technology, the utilized printers make use of a thermoplastic type of filament and this filament is heated to attend its melting temperature, followed by layer-to-layer extrusion, resulting in the construction of a three-dimensional structure. Printing pieces in a tiered manner ensures continuous availability of the specified material. The filament is brought to a semi-liquid condition by the heating element in the liquefier head, after which it is extruded via the nozzle to the printing area where the actual component is printed. The most crucial step in this method is to utilize the subsequent layer before it turns solid, as solidification before to fusing might have an additional impact on the other building part character traits. There are definitely some benefits and drawbacks to the FDM process. The FDM process offers several advantages, including operational simplicity, lower production costs, and relatively faster printing speeds; however, it also presents limitations such as reduced resolution, limited material compatibility, and the potential for structural weaknesses at interlayer boundaries.

6.7.1 FDM PROCESS PARAMETERS

1. **Nozzle diameter:** Road width is significantly impacted by the nozzle's diameter because it directly affects the pressure drop within the liquefier; evidence from experiments also shows that the nozzle's width to diameter ratio, or L/D, additionally plays a role in the differences in the pressure drop.
2. **Build orientation:** It explains how the specified component is modified on the build platform with respect to the machine tool's three main axes, X, Y, and Z.
3. **Layer height:** This is the quantity of material that is deposited in a single pass along the FDM machine's vertical axis. Substances are constantly deposited at heights lower than the extruder's nozzle diameter.
4. **Raster width:** It indicates the breadth of the deposition route utilized to construct the specified element. The nozzle diameter has a big impact on the route width.

5. **Air gap:** It signifies the space that exists between two adjacent bead deposits. It can occur for the air gap value to be negative, positive, or zero. The materials that have been deposited are in direct contact with one another when there are no air spaces. Positive air gaps allow for the material to settle in progressively broken away, creating a loosely assembled structure where quick assembly of the specified component is crucial. When a denser structure is required, a negative air gap is utilized, and when there is no time restriction. Beads partially occupy the same space in the negative air gap, creating a denser composite.
6. **Infill density:** The volume of material printed on a particular component is indicated by the infill density. The printed component's qualities are directly influenced by the infill density. While the denser component has better mechanical qualities than the earlier one, but requires more time to create, the lower density has a significant impact on the mechanical properties [49, 50].

One distinct nozzle for printing polymer material and another for producing the fiber-polymer material are housed in the single print head of this FDM printer. There is no contact between the fiber-polymer and polymer filaments, and the print head can only run one nozzle at a time. Using a filament-by-filament, layers by layers method, the FDM process comprises four sequential processes that are carried out continuously to print a single FDM fiber-polymer filament, that serves as the fundamental "building block" of a 3D printed laminate [51].

Furthermore, vibration and surrounding circumstances are crucial for 3D printing. It was recommended to use a 3D printer from the Rigid2 Zero brand, which comes in a closed box with a full metal body, to avoid the drawbacks of these conditions. One of the industry's most favored engineering polymers is PA12 material. The PA12 filament material from ESUN brand was chosen because of its superior mechanical and chemical characteristics [52].

Multiple steps are involved to print a device using FDM technology which is dependent on the complexity of the printable device. The following steps are one such example of the process:

Device design: A design is created as per the requirement of the device. Then, it is validated using digital models matched to a patient's anatomy.

Software workflow: The design is converted to buildable file. It will be then sent to the FDM printer. In printer, the file will often divide the design into multiple layers, including additional supportive material and guide the printer on how to build the device on the printer platform. During software set up, settings for the material, type of design, and intention of usage are required to change as per requirement.

Material controls: The manufacturing process of 3D printing needs high quality materials which meet a specific quality to build devices with consistent high quality. To ensure requirements and agreements is called material controls which must be scanned and tested for every batch of material. It is established among the suppliers, buyers, and end-users of the material.

Printing: Printing is carried out according to the instructed file.

Post-processing: After printing completion, cleaning of residual debris, controlled cooling (also called annealing), and/or additional steps like drilling, cutting, polishing, and sterilization are carried out as per requirement.

Process validation and verification: After post-processing, characterization and checking of some parameter of the printed device are subjected to be checked whether they meet the specification or not. Manufacturers must validate their production processes before starting the production. Process validation guarantees that a manufacturing process will produce a quality product which will meet specification.

Testing: Quality control method and results are submitted to the regulatory authority for checking whether the device meets regulatory requirements or not. They also check whether it is reasonably safe and effective for its intended use or not.

6.8 FUTURE PROSPECTIVES

Several factors and material parameters are needed to be considered when part is designed for FDM production and many issues are still needed to be overcome. One of the drawbacks that affects FDM process in industry is the anisotropy of the manufactured parts which is mostly due to weakness given by the imperfect adherence among layers. However, this obstacle can be overcome by proper manufacturing engineering to better replicate the complex architecture of biological tissues since most biological tissues are anisotropic in nature [53].

REFERENCES

1. Tim. (2021, Jan 18). *Brief history of medical devices and regulations*. Winovia. https://www.winovia.com/brief-history-of-medical-devices-and-regulations/
2. Surgical Instruments from Ancient Rome. (2024). Ancient roman surgical instruments. Retrieved April 17, 2024, from http://exhibits.hsl.virginia.edu/romansurgical/
3. Introduction to Medical Devices and their regulatory framework: 2. History of medical devices | *EUPATI Open Classroom*. (2024). Retrieved April 17, 2024, from https://learning.eupati.eu/mod/page/view.php?id=927&forceview=1
4. Center for Devices and Radiological Health. (2024, April 12). *Medical devices*. FDA. https://www.fda.gov/medical-devices
5. About us | Epo.org. (2024). Retrieved April 17, 2024, from https://www.epo.org/en/about-us
6. Definitions for Personalized Medical Devices. (2024). https://www.imdrf.org/working-groups/personalized-medical-devices
7. Vrana, N. E., Palm, K., & Lavalle, P. (2020). Personalization of medical device interfaces: Decreasing implant-related complications by modular coatings and immunoprofiling. *Future Science OA*, *6*(8), FSO607. https://doi.org/10.2144/fsoa-2020-0074
8. Aimar, A., Palermo, A., & Innocenti, B. (2019). The role of 3D printing in medical applications: A state of the art. *Journal of Healthcare Engineering*, *2019*, 1–10. https://doi.org/10.1155/2019/5340616
9. Awad, A., Gaisford, S., & Basit, A. W. (2018). Fused deposition modelling: Advances in engineering and medicine. In A. W. Basit and S. Gaisford (Eds.), *3D printing of pharmaceuticals* (Vol. 31, pp. 107–132). Springer International Publishing. https://doi.org/10.1007/978-3-319-90755-0_6
10. Long, J., Gholizadeh, H., Lu, J., Bunt, C., & Seyfoddin, A. (2017). Application of fused deposition modelling (FDM) method of 3D printing in drug delivery. *Current Pharmaceutical Design, 23*(3), 433–439. https://doi.org/10.2174/1381612822666161026162707
11. Sai, P. C., & Yeole, S. N. (2014). *Fused deposition modeling—Insights.*
12. Classification of Products as Drugs and Devices and Additional Product Classification Issues | FDA. (2024). Retrieved April 17, 2024, from https://www.fda.gov/regulatory-information/search-fda-guidance-documents/classification-products-drugs-and-devices-and-additional-product-classification-issues
13. Medical devices. (2024). Retrieved April 18, 2024, from https://www.who.int/health-topics/medical-devices
14. Gelijns, A. C., et al. (1989). 3: The development of medical devices. In *Technological innovation: Comparing development of drugs, devices, and procedures in medicine*. National Academies Press. https://www.ncbi.nlm.nih.gov/books/NBK222708/
15. Why Is 3D Printing for the Medical Device Industry Important? (2024). Retrieved April 17, 2024, from https://www.enggtechnique.com/resource/blog-detail/why-is-3d-printing-for-the-medical-device-industry-important
16. Zhou, H., & Bhaduri, S. B. (2019). 3D printing in the research and development of medical devices. In *Biomaterials in translational medicine* (pp. 269–289). Elsevier. https://doi.org/10.1016/B978-0-12-813477-1.00012-8

17. Tsoulfas, G., Bangeas, P. I., Suri, J. S., & Papadopoulos, V. N. (2020). Introduction: The role of 3D printing in surgery. In *3D printing: Applications in medicine and surgery* (pp. 1–6). Elsevier. https://doi.org/10.1016/B978-0-323-66164-5.00001-5
18. Chen, Y., Bian, L., Zhou, H., Wu, D., Xu, J., Gu, C., Fan, X., Liu, Z., Zou, J., Xia, J., & Xu, Z. (2020). Usefulness of three-dimensional printing of superior mesenteric vessels in right hemicolon cancer surgery. *Scientific Reports, 10*(1), 11660. https://doi.org/10.1038/s41598-020-68578-y
19. Bianchi, G., Frisoni, T., Spazzoli, B., Lucchese, A., & Donati, D. (2021). Computer assisted surgery and 3D printing in orthopaedic oncology: A lesson learned by Cranio-Maxillo-facial surgery. *Applied Sciences, 11*(18), 8584. https://doi.org/10.3390/app11188584
20. Grygier, D., Kujawa, M., & Kowalewski, P. (2022). Deposition of biocompatible polymers by 3D printing (FDM) on titanium alloy. *Polymers, 14*(2), 235. https://doi.org/10.3390/polym14020235
21. Sharma, A., & Rai, A. (2022). Fused deposition modelling (FDM) based 3D & 4D printing: A state of art review. *Materials Today: Proceedings, 62*, 367–372. https://doi.org/10.1016/j.matpr.2022.03.679
22. Wang, X., Jiang, M., Zhou, Z., Gou, J., & Hui, D. (2017). 3D printing of polymer matrix composites: A review and prospective. *Composites Part B: Engineering, 110*, 442–458. https://doi.org/10.1016/j.compositesb.2016.11.034
23. Serrano-Aroca, Á., & Llorens-Gámez, M. (2017). Dynamic mechanical analysis and water vapour sorption of highly porous poly(methyl methacrylate). *Polymer, 125*, 58–65. https://doi.org/10.1016/j.polymer.2017.07.075
24. Serra, T., Planell, J. A., & Navarro, M. (2013). High-resolution PLA-based composite scaffolds via 3-D printing technology. *Acta Biomaterialia, 9*(3), 5521–5530. https://doi.org/10.1016/j.actbio.2012.10.041
25. 3D printing for chemical, pharmaceutical and biological applications | *Nature Reviews Chemistry*. (2024). Retrieved April 18, 2024, from https://www.nature.com/articles/s41570-018-0058-y
26. 3D bioprinting of tissues and organs | Nature Biotechnology. (2024). Retrieved April 17, 2024, from https://www.nature.com/articles/nbt.2958
27. Abreu, E. (2006). Peter Ma X, Elisseeff Jennifer (Editors): Scaffolding in tissue engineering. CRC Press, Taylor and Francis Group. xvi+638 pages, ISBN 1-57444-521-9 (Hardcover). *BioMedical Engineering OnLine, 5*(1), 52, 1475-925X-5–52. https://doi.org/10.1186/1475-925X-5-52
28. Shaqour, B., Samaro, A., Verleije, B., Beyers, K., Vervaet, C., & Cos, P. (2020). Production of drug delivery systems using fused filament fabrication: A systematic review. *Pharmaceutics, 12*(6), 517. https://doi.org/10.3390/pharmaceutics12060517
29. Youssef, A. A. A., Cai, C., Dudhipala, N., & Majumdar, S. (2021). Design of topical ocular ciprofloxacin nanoemulsion for the management of bacterial keratitis. *Pharmaceuticals, 14*(3), 210. https://doi.org/10.3390/ph14030210
30. Goyanes, A., Buanz, A. B. M., Hatton, G. B., Gaisford, S., & Basit, A. W. (2015). 3D printing of modified-release aminosalicylate (4-ASA and 5-ASA) tablets. *European Journal of Pharmaceutics and Biopharmaceutics, 89*, 157–162. https://doi.org/10.1016/j.ejpb.2014.12.003

31. Beck, R. C. R., Chaves, P. S., Goyanes, A., Vukosavljevic, B., Buanz, A., Windbergs, M., Basit, A. W., & Gaisford, S. (2017). 3D printed tablets loaded with polymeric nanocapsules: An innovative approach to produce customized drug delivery systems. *International Journal of Pharmaceutics, 528*(1–2), 268–279. https://doi.org/10.1016/j.ijpharm.2017.05.074
32. Khorsandi, D., Fahimipour, A., Abasian, P., Saber, S. S., Seyedi, M., Ghanavati, S., Ahmad, A., De Stephanis, A. A., Taghavinezhaddilami, F., Leonova, A., Mohammadinejad, R., Shabani, M., Mazzolai, B., Mattoli, V., Tay, F. R., & Makvandi, P. (2021). 3D and 4D printing in dentistry and maxillofacial surgery: Printing techniques, materials, and applications. *Acta Biomaterialia*, *122*, 26–49. https://doi.org/10.1016/j.actbio.2020.12.044
33. Dawood, A., Marti, B. M., Sauret-Jackson, V., & Darwood, A. (2015). 3D printing in dentistry. *British Dental Journal*, *219*(11), 521–529. https://doi.org/10.1038/sj.bdj.2015.914
34. Health Canada. (2004, July 26). *Medical devices* [Navigation page]. https://www.canada.ca/en/health-canada/services/drugs-health-products/medical-devices.html
35. *European Medicines Agency*. (2024). Retrieved April 19, 2024, from https://www.ema.europa.eu/en/medical-device
36. Pharmaceuticals and Medical Devices Agency. (2024). Pharmaceuticals and medical devices agency. Retrieved April 19, 2024, from https://www.pmda.go.jp/english/index.html
37. Medical devices reforms | Therapeutic Goods Administration (TGA). (2024). Retrieved April 19, 2024, from https://www.tga.gov.au/how-we-regulate/supply-therapeutic-good/supply-medical-device/medical-devices-reforms
38. National Medical Products Administration. (2024). Retrieved April 19, 2024, from https://english.nmpa.gov.cn/
39. Medical device & diagnostics. (2024). Retrieved April 19, 2024, from https://cdsco.gov.in/opencms/opencms/en/Medical-Device-Diagnostics/Medical-Device-Diagnostics/
40. International Medical Device Regulators Forum (IMDRF) | *International Medical Device Regulators Forum*. (2024). Retrieved April 19, 2024, from https://www.imdrf.org/
41. Cailleaux, S., Sanchez-Ballester, N. M., Gueche, Y. A., Bataille, B., & Soulairol, I. (2021). Fused deposition modeling (FDM), the new asset for the production of tailored medicines. *Journal of Controlled Release, 330*, 821–841. https://doi.org/10.1016/j.jconrel.2020.10.056
42. Pavan Kalyan, B., & Kumar, L. (2022). 3D printing: Applications in tissue engineering, medical devices, and drug delivery. *AAPS PharmSciTech, 23*(4), 92. https://doi.org/10.1208/s12249-022-02242-8
43. Redaelli, D. F., Abbate, V., Storm, F. A., Ronca, A., Sorrentino, A., de Capitani, C., Biffi, E., Ambrosio, L., Colombo, G., & Fraschini, P. (2020). 3D printing orthopedic scoliosis braces: A test comparing FDM with thermoforming. *International Journal of Advanced Manufacturing Technology, 111*. https://doi.org/10.1007/s00170-020-06181-1
44. Thurzo, A., Kočiš, F., Novák, B., Czako, L., & Varga, I. (2021). Three-Dimensional Modeling and 3D Printing of Biocompatible Orthodontic Power-Arm Design with

Clinical Application. *Applied Sciences, 11*(20), 9693. https://doi.org/10.3390/app11209693
45. Ventola, C. L. (2024). *Medical applications for 3D printing: Current and projected uses*. D Printing.
46. Patel, R., Desai, C., Kushwah, S., & Mangrola, M. H. (2022). A review article on FDM process parameters in 3D printing for composite materials. *Materials Today: Proceedings, 60*, 2162–2166. https://doi.org/10.1016/j.matpr.2022.02.385
47. Singh, D., Singh, R., & Boparai, K. S. (2018). Development and surface improvement of FDM pattern based investment casting of biomedical implants: A state of art review. *Journal of Manufacturing Processes, 31*, 80–95. https://doi.org/10.1016/j.jmapro.2017.10.026
48. Ali, M. A., Rajabi, M., & Sudhir Sali, S. (2020). Additive manufacturing potential for medical devices and technology. *Current Opinion in Chemical Engineering, 28*, 127–133. https://doi.org/10.1016/j.coche.2020.05.001
49. Jaisingh Sheoran, A., & Kumar, H. (2020). Fused deposition modeling process parameters optimization and effect on mechanical properties and part quality: Review and reflection on present research. *Materials Today: Proceedings, 21*, 1659–1672. https://doi.org/10.1016/j.matpr.2019.11.296
50. Solomon, I. J., Sevvel, P., & Gunasekaran, J. (2021). A review on the various processing parameters in FDM. *Materials Today: Proceedings, 37*, 509–514. https://doi.org/10.1016/j.matpr.2020.05.484
51. Hu, Y., Ladani, R. B., Brandt, M., Li, Y., & Mouritz, A. P. (2021). Carbon fibre damage during 3D printing of polymer matrix laminates using the FDM process. *Materials & Design, 205*, 109679. https://doi.org/10.1016/j.matdes.2021.109679
52. Kam, M., İpekçi, A., & Şengül, Ö. (2023). Investigation of the effect of FDM process parameters on mechanical properties of 3D printed PA12 samples using Taguchi method. *Journal of Thermoplastic Composite Materials, 36*(1), 307–325. https://doi.org/10.1177/08927057211006459
53. Grivet-Brancot, A., Boffito, M., & Ciardelli, G. (2022). Use of polyesters in fused deposition modeling for biomedical applications. *Macromolecular Bioscience, 22*(10), 2200039. https://doi.org/10.1002/mabi.202200039
54. Buj-Corral, I., Tejo-Otero, A., & Fenollosa-Artés, F. (2021). Use of FDM technology in healthcare applications: Recent advances. In H. K. Dave & J. P. Davim (Eds.), *Fused deposition modeling based 3D printing* (pp. 277–297). Springer International Publishing. https://doi.org/10.1007/978-3-030-68024-4_15

while maintaining adherence to quality and regulatory benchmarks. Its primary focus is to examine the present regulatory landscape governing large-scale 3D printing in pharmaceuticals, while also addressing the existing gaps and obstacles that must be recognized and addressed to propel the contemporary production of medicines forward.

7.1 INTRODUCTION

The emergence of 3D printing, also known as additive manufacturing (AM), has brought about a significant transformation in the field of medical device innovation and personalized healthcare. As the aging population in Europe is projected to increase to 35% over the next three decades, there is a pressing need for innovative medical solutions to detect and address health issues early, thereby reducing hospital stays and empowering patients to better monitor their health [1]. The COVID-19 pandemic has further underscored the demand for medical devices, particularly those that support sustainable homecare for chronic conditions [2]. 3D bioprinting has emerged as a promising technology for creating customized medical devices, including bone implants, tissue-engineered scaffolds, personalized drug delivery systems, biosensors, and other digital healthcare tools [3]. Biomedical companies must adapt to this changing landscape by focusing on consistent product development and tailoring solutions to meet the needs of patients, especially those with chronic illnesses. Regulatory agencies like the European Medicines Agency (EMA) advocate for leveraging digital tools and data to enhance pharmaceutical development and integration with scientific advancements [4, 5]. However, designing medical devices presents challenges, especially with the implementation of new regulations such as the Medical Device Regulation (MDR) in the EU. Developing medical devices requires intricate collaboration among regulatory bodies, diverse user groups, and interdisciplinary sciences. To navigate this complexity, both the US Food and Drug Administration (USFDA) and the EMA emphasize the adaption of quality by design (QbD) principles in quality management frameworks. By integrating QbD principles early in the development process, knowledge transfer is facilitated, and a continuous lifecycle management approach is established, thereby promoting the acceptance of 3D printing technology and ensuring regulatory compliance [6, 7]. In the rapidly evolving landscape of pharmaceuticals, 3D printing has emerged as a transformative technology with immense potential.

However, alongside its promises come regulatory challenges that demand careful consideration and navigation. This chapter delves into the regulatory framework surrounding 3D-printed pharmaceuticals, unraveling complexities and offering insights into compliance strategies.

7.2 TYPES OF 3DP

The 3D printing industry has experienced substantial growth by introducing new technologies and terminology. 3D printing serves as a nexus between art and science, facilitating the swift and accurate fabrication of intricate structures and components. The techniques covered by the ASTM categorization system include deposition, binding, and polymerization. These techniques are used in a variety of sectors, including drug delivery systems, medical devices, and biomedical applications [8]. Medical uses span a wide array, encompassing medical models, implants, tools, aids, prostheses, and biomanufacturing. In 2012, ASTM International established a standard terminology, categorizing AM technologies into seven overarching groups.

7.2.1 BINDER JETTING

Binder jetting is a printing method employing a binder and powdered material to construct a self-supporting print on a powder bed. In this process, the building material exists in powder form, with the binder serving as an adhesive between layers. This technique enables color printing and utilizes various polymers as well as materials like stainless steel and glass. Unlike its utility for larger parts, it's not typically employed for tools, instruments, or components for medical devices. The procedure involves spreading a thin layer of powder on the build platform, followed by the application of a binding solution to specific areas of the digital design. Afterward, excess powder is eliminated and preserved for future use. Although binder jetting can accommodate larger parts, it tends to be more costly for larger systems [9–11].

7.2.2 DIRECTED ENERGY DEPOSITION

It is a sophisticated printing technique utilized for either repairing or augmenting material onto existing components. It employs materials in either powder or wire form, along with a multi-axis arm to heat and during printing [27]. Upon completion, the sculpture must be extracted from the resin, and any excess resin must be drained. The first type of 3D printing, known as photopolymerization, operates by solidifying layers of resin through a chemical reaction induced by specific lighting, typically from a laser or projector [28]. While this method can produce precise pieces with sharp edges, it is relatively expensive and offers a limited selection of materials.

7.3 PHARMACEUTICAL APPLICATION OF 3DP

7.3.1 EXPLORING THE IMPACT OF 3DP TECHNOLOGY ON DOSAGE FORM DESIGNING

7.3.1.1 TABLET

Various 3D printing (3DP) technologies, including injection molding (IJ), free direct dyeing (FDM), and solid-state lasers (SLS), are currently utilized in the manufacturing of pharmaceutical dosage forms. Among pharmaceutical products, tablets are highly favored, prompting extensive exploration of 3DP technologies for pharmaceutical products designing and production. Tablets can be broadly classified into single API tablets and multiple API tablets [29]. Initially, single API tablets were primarily used to produce simple immediate release (IR) tablets, with the FDM method favored for its straightforward fabrication procedures. Drug release profiles are impacted by the physical characteristics of the final tablets, which are greatly influenced by the selection of 3DP materials and processes [30]. Additionally, the proportions of formulation components play a crucial role in shaping the physical attributes of tablets, ultimately altering drug release behaviors. The versatility of 3DP technology extends to various types of modified release tablets, such as enteric tablets, gastro-floating tablets, and drug-laden solid dosage forms with complex shapes and structures [31]. Techniques such as SLA (stereolithography) and IJ (inkjet) are also suitable for the production of tablets, with SLA successfully producing extended-release (ER) tablets of substances like 4-aminosalicylic acid or paracetamol using specific composition ratios of various polumers [32].

As of right now, the most promising methods for producing oral solid dosage forms are IJ and FDM techniques. Although FDM shows promise in creating complicated geometries for modified release dosage forms, its use is somewhat constrained by the availability of polymer excipients [33]. One potential solution to this limitation lies in polymer blends, which can be adjusted to enhance miscibility between excipients, solubility in dissolution media, and fusion during the FDM process [34]. Recent research demonstrates how 3DP technology is used to produce polypills with controlled release patterns. For example, Khaled et al. made a polypill with glipizide, nifedipine, and captopril specifically for individuals with hypertension and diabetes using 3D extrusion-based printing [35]. This polypill included sustained release compartments for glipizide and nifedipine, a connecting layer, and an osmotic pump compartment for captopril. Additionally, a polypill comprising five APIs was made using 3DP technology. It included two compartments with separate controlled release profiles—one for prolonged release and the other for quick release [35].

7.3.1.2 IMPLANTS

3D printed implants (3DP) represent a promising treatment modality for chronic ailments, offering prolonged drug release and enhanced localized delivery. Recent advancements in 3DP-based implants feature intricate micro- and macro-structures, enabling the loading of multiple active pharmaceutical ingredients (APIs) and the development of enhanced drug release patterns [36]. For instance, Huang et al. devised a levofloxacin implant with a bimodal release pattern, exhibiting both pulsatile and sustained drug release [37]. Apart from their use in medication administration, 3DP technology is also utilized in the creation of scaffolds for bone regrowth, in which several APIs are included into the implants to provide extra therapeutic advantages [38]. Wu et al. (2009) developed a multi-drug implant designed for the management of bone tuberculosis. This implant was engineered to deliver isoniazid and rifampicin in a predetermined sequence within each layer using 3DP technology [39]. Wu et al. (2016), in another study, employed 3DP technology to create a multi-drug implant for addressing chronic osteomyelitis. This implant featured levofloxacin (LVFX) and tobramycin (TOB) as the APIs [40]. Moreover, 3DP has made it easier to create patches-like implants that are designed to deliver drugs over a long period of time for chronic conditions like pancreatic cancer. Yi et al. (2016) used 3DP to create an implant that allows for site-specific medication delivery by targeting particular tissues or organs [41]. This biodegradable implant patch demonstrated notable efficacy in adhering to the pancreas in a murine model, highlighting the promise of 3DP-based biodegradable implants for efficient localized administration of anticancer drugs.

7.3.2 PERSONALIZED MEDICINE

A major difficulty is ensuring the safety and effectiveness of pharmaceuticals for particular populations, such as children and the elderly. Children react differently to drugs due to their unique sensitivity, while the elderly often experience reduced absorption and metabolism rates. Standard drug dosages may not always be suitable and manually administering medication to children can lead to unwanted side effects [42]. 3DP technology presents a promising strategy by allowing for the customization of medications based on individual needs. This technology enables the production of tailored medicines by adjusting various parameters [43]. Specifically, for pediatric patients, 3DP facilitates the development of personalized and low-dose medications that are not only more palatable but also enhance adherence to treatment regimens [44]. Additionally, for elderly individuals who struggle with swallowing, 3DP can generate porous formulations that are easier to ingest and it can even combine multiple medications into a single tablet for convenience [45]. Furthermore, 3DP allows for the creation of specially shaped medications or symbols to assist patients with visual impairments. This innovative approach not only enhances medication adherence but also addresses specific patient needs. Companies like FabRx are at the forefront of developing and implementing 3DP technology to advance personalized medicine, aiming to improve patient outcomes and overall healthcare delivery.

7.3.3 OPTIMIZING DRUG DELIVERY: PRECISION IN RELEASE MECHANISMS

Tablets constitute the majority of solid oral dosage forms, making up 70% of all produced dosage forms. While traditional manufacturing methods are cost-effective, they have limitations in developing preparations tailored to individual needs [46]. Controlled-release formulations offer precise control over drug release, thereby minimizing side effects and enhancing effectiveness. However, there are many obstacles for conventional production methods to overcome in this area [47]. One very adaptable substitute is three-dimensional (3D) printing technology, which is ideal for assembling medications, building models, and modifying specifications to create complex preparations [48]. Triastek's 3D printed product, T19, exemplifies this advancement. T19 is a controlled-release formulation specifically designed to align with the circadian rhythm of rheumatoid arthritis [49]. By maintaining optimal blood concentration during the day, T19 provides improved therapeutic options for patients experiencing severe symptoms. This innovative approach not only enhances treatment efficacy but also expands the possibilities for personalized medication solutions.

7.3.4 EFFICIENT INCORPORATION OF MANUFACTURING PROCESSES

Pharmaceutical companies often face challenges with their high-capacity production setups and bulky equipment, which lack the agility needed to swiftly accommodate changes in drug types. 3DP technology emerges as a solution by seamlessly blending rapid manufacturing capabilities with compact machinery, streamlined production workflows, automated digital processes, and the flexibility to adapt to varying drug types [50]. SSE technology, for example, makes it possible to directly replace single-use syringes that contain various drug formulations. Moreover, 3DP proves advantageous for small-scale drug manufacturing due to its lower costs for producing small batches and its ability to integrate manufacturing processes efficiently, making it particularly suitable for scenarios with limited time and resources. Merck, for instance, has harnessed 3DP to expedite clinical trials, anticipating a 60% decrease in preparation development time and a 50% reduction in the API required for medication preparation across clinical phases I to III [51].

7.4 REGULATORY GUIDELINES

7.4.1 THE CURRENT SCENARIO FOR MASS PRODUCTION OF PHARMACEUTICALS

The current regulatory framework for 3DP is overseen by the US Federal Regulations, mandating a chemistry, manufacturing, and control (CMC) standard for both novel and generic drugs. The FDA offers three main pathways for new drugs: 505(b)(1), 505(b)(2), and the sole pathway for generic drugs, 505(j) [52]. These routes provide a basic framework for both novel and generic pharmaceuticals, and they are sufficiently flexible to accommodate evolving technology.The ICH was founded in 1990 with the goal of bringing the scientific and technological underpinnings of drug registration into global agreement. It is possible to consistently apply the four ICH topics—quality, safety, efficacy, and the multidisciplinary guideline—to 3DP technology for pharmaceutical products [53]. The ICH advocates for the advancement of regulatory science and innovation to promote cleaner, enabling more adaptable and streamlined production of pharmaceutical goods.

For a novel medicine whose active components have not been approved before, the standard novel medicine application (NDA) is used to get approval. A novel molecule's safety and effectiveness for the specified indication must be established through this pathway, which calls for comprehensive clinical and non-clinical research. A 505(b)(2) application, on the other hand, is based on investigations that were not carried out by or on behalf of the applicant, and for which the applicant has not received permission to use or reference the results [54].

The EMA provides detailed guidelines for a range of topics, including active ingredients, manufacturing, impurities, specifications, excipients, packaging, stability, pharmaceutical development, particular product types, and lifecycle management. It also oversees the regulatory environment for pharmaceutical products in areas outside of the US. With additional qualities unique to 3D printed items, these quality criteria may be applied to any medicinal product [55]. Pharmaceutical items are governed by pharmacopoeias and laws specific to other areas, such as Asia and South America. For mass manufacturing and distribution inside their individual nations, 3DP ought to come under their regulatory purview.

7.4.2 CONSIDERATIONS FOR 3D PRINTING IN PHARMACEUTICAL MANUFACTURING

7.4.2.1 PRODUCT DESIGN AND PROCESS CONSIDERATIONS

Manufacturing pharmaceuticals is revolutionized by 3DP, which makes it possible to produce 3D items layer by layer from digital blueprints. This method builds objects sequentially, layer by layer, joining cach to the layer below through controlled movements of the print head or base plate [56]. The versatility of 3DP allows pharmaceutical companies to create personalized dosage forms, including complex drug delivery systems, resulting in better patient compliance, minimized side effects, and improved treatment efficacy.

Various 3DP platforms exist, each with its own input materials and operating principles, closely tied to the type of 3D printer utilized. Specific product designs, like Spritam, are tailored to certain 3D printers, such as those employing ZipDose technology and IJ 3DP [57]. In contrast, traditional pharmaceutical processes rely on factors like dosage form and manufacturing method, with tablets commonly produced through compression or molding, and capsules filled with powders or granules [58]. The decision between 3DP and conventional techniques depends on a number of variables, including product complexity, production speed, and cost.

While 3DP can replicate established pharmaceutical dosage forms, it excels in creating intricate geometries, including spherical, cylindrical, pyramidal, and multi-layered combinations. Such complex designs can impact drug release mechanisms, facilitating bimodal drug release mechanisms that conventional methods struggle to achieve. Recent advancements demonstrate the precise control of drug and excipient arrangement in multiple layers of solid oral dosage forms [58, 59].

Crucial considerations in 3D printed pharmaceutical development include the intended patient population, whether individual or group-based, as well as factors like molecule type, dosage precision, delivery route, and quality attributes. Manufacturing processes, regulatory requirements, and cost-effectiveness also influence product development. Evaluating these factors ensures successful product development. Pharmaceutical companies select 3DP platforms based on product design, with considerations for material compatibility, printing speed, and accuracy. The chosen platform must meet product requirements while adhering to regulatory standards and cost considerations. Careful evaluation of these factors ensures the successful utilization of 3DP in pharmaceutical industry [60, 61].

Pharmaceutical companies can employ a range of 3DP platforms, contingent upon the computer-aided design-generated internal, external, and functional design of the product. While there are many various ways that 3D printers function, most of them adhere to a schematic like the one shown in Figure 7.1 and the sequential steps listed below for manufacturing pharmaceuticals:

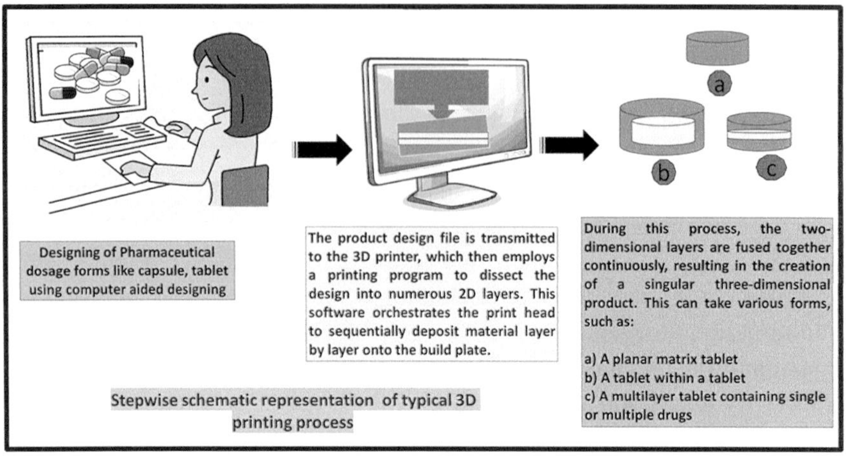

FIGURE 7.1 A step-by-step graphic illustration of a standard 3D printing procedure.

- *Initial creation of digital design:* CAD software is used to produce the 3D digital design of the intended product.
- *Conversion and slicing:* The design is transformed into a machine-readable file format that describes the 3D model's exterior. Then, printing programs "slice" these surfaces into separate printable 2D layers, sending the printer digital instructions layer by layer. During printing, the program decides where to print scaffolding support material and moves the print head.
- *Material processing:* At this point, raw materials are treated. Materials can be processed into granules, filaments, dry mixed powder, or binder solutions, depending on the printing technique, to facilitate printing. The print fluid or granules/powder may include active medication ingredients, sometimes in combination.
- *Layer-by-layer solidification:* The build-up cycle is when raw components are automatically added and consolidated layer by layer to create the desired result.
- *Post-printing treatments:* Products could need to undergo other processes like drying, sintering, or polishing after printing. Unprinted materials can be collected and recycled for future use; however, if they include active components, more characterization may be necessary to determine whether to separate or degrade the medicine [62].

When a certain printer type is chosen, a number of process concerns usually come into play which are as follows:

a. Feasibility for large-scale production
b. Obstacles in transitioning from research and development to large-scale production
c. Production rate and market needs
d. Ensuring uniform replication of individual doses and adaptability
e. Integration of progressive pharmaceutical development, manufacturing, and quality assurance models (process analytical technology)
f. Adherence to qualification and validation standards within a current good manufacturing practice (cGMP) setting
g. Process management through software
h. Software validation process

7.4.2.2 QUALITY CONTROL IN 3D PRINTING PHARMACEUTICALS

In the pharmaceutical field, integrating 3DP demands a meticulous level of control to uphold consistent product quality within a standard commercial framework. Given its relative novelty in mass pharmaceutical production, navigating potential challenges like sudden software glitches and unanticipated shifts in process variables is paramount for ensuring uniformity and excellence in product output. Precision in equipment design, product alignment, layer thickness, print head elevation, printing velocity, and patterns is fundamental across all 3DP procedures. Therefore, establishing robust quality control protocols becomes imperative to continuously monitor and fine-tune these variables throughout the printing continuum. Moreover, regular upkeep and calibration of equipment play a pivotal role in forestalling deviations that could jeopardize product integrity [63].

While 3DP manufacturing might not match the swiftness of traditional pharmaceutical manufacturing methods, it necessitates the delineation of appropriate process parameters or thresholds during its developmental phase. The capability to promptly detect process disturbances and enact corrective measures, such as employing feedback control mechanisms, further fortifies the consistency of product quality [64]. Quality frameworks must be adept at gauging measurements at intervals commensurate with the specific type of 3DP employed and the duration of the manufacturing cycle. This assessment should account for inherent process risks, documented process variances, and the requisite duration for processing in the equipment. A comprehensive risk assessment should be undertaken to comprehensively grasp and effectively manage any potential impact on product quality across all operational scenarios. Furthermore, early detection of aberrations and probable defect prevention are facilitated by real-time monitoring of crucial process parameters. Instituting a robust quality management regime, inclusive of regular audits and evaluations, can pinpoint areas for enhancement and ensure sustained quality control.

As with any customary commercial pharmaceutical manufacturing endeavor, establishing fitting sampling, testing, quality control methodologies, and equipment protocols is essential for identifying and rejecting materials that fall outside specified parameters. It's equally crucial to delineate clear standards for acceptable quality levels and institute mechanisms for handling deviations from these benchmarks. Routine audits and inspections serve to validate compliance with regulatory mandates and industry norms, upholding adherence to the highest standards of quality assurance [65, 66].

7.4.2.3 RAW MATERIAL AND INTERMEDIATE CONTROL

3DP operates through a meticulous layer-by-layer procedure, demanding heightened oversight of raw materials due to the profound physical and chemical alterations they undergo. The characteristics of the final drug product are greatly influenced by the flow properties and micromeritics of materials introduced during the printing process. In powder bed IJ 3DP, inadequate characterization of these properties can lead to flawed layer adhesion, resulting in irregular layer thicknesses and divergent product designs [25, 67].

The choice of APIs and excipients hinges on their suitability for printing and stability under various conditions, including the specific 3DP method employed. Sustaining a close rapport with raw material suppliers and vigilant monitoring of batch-to-batch variability are essential. This entails creating exacting requirements for incoming materials and using standards and methodologies for verified analytical testing. Keeping a tight contact with raw material suppliers and keeping a constant eye on lot-to-lot fluctuation are essential. To ensure quality, it's essential to establish appropriate specifications for materials and use validated analytical testing techniques. Common characteristics of raw materials include micromeritics, flow, viscoelastic properties, thermal characteristics, viscosity, presence of impurities, and solid state and moisture content. For polymeric materials, additional attributes like monomer composition, purity, molecular weight distribution, and thermal attributes are crucial.

Various intermediate materials may be integral to the 3DP process, depending on the printer type. For example, powder IJ printing often involves two primary intermediates: a powder blend and a print fluid. Thorough characterization of these intermediates is vital for subsequent extrusion-based printing steps. Control strategies for raw materials and intermediates should be rooted in a comprehensive understanding of product and process requirements and may involve process analytical technology (PAT) monitoring. PAT monitoring allows for real-time analysis of critical process parameters, ensuring consistency and quality in the final product. By implementing control strategies based on thorough characterization and monitoring, manufacturers can optimize their 3DP processes for efficiency and reliability [22, 68].

7.4.2.4 PROCESS EQUIPMENT CONTROL

Equipment control plays a pivotal role in 3DP, given the diverse array of printers available, each operating on distinct principles. Among the most prevalent are powder IJ and FDM printers. Both rely on a software workflow, making software control indispensable for their functionality. To ensure effective software control, careful attention must be paid to software attributes throughout development and commercial manufacturing. This entails assessing how software interacts with drug product design, understanding clinically relevant design parameters, and managing complex design files with precision. File format conversions must be executed meticulously, with the final product design archived and maintained in accordance with the AM file format (AMF). Furthermore, regular software updates and maintenance are essential to uphold optimal performance and compatibility with emerging technologies. Clear protocols for software validation and verification should also be established for precision and ensure reliability of the process [63, 69].

Common manufacturing issues in 3DP encompass variations in print droplet formation, inadequate drying, fluctuations in harvesting, and concerns related to binder or fluid viscosity, surface tension, and dynamic viscosity [70]. These risks can result in tablet disintegration, microbial contamination, inconsistencies in content and weight uniformity, impurities causing degradation, polymorphic binding, layer agglomeration, and delamination. Parameters including feed rate, temperature, jetting rate, print head and powder bed speed, print area density, print fluid fill height, and distance between the print head and powder bed must all be monitored in real time for effective process management. It is also essential to assess mix homogeneity using PAT or near-infrared (NIR) spectroscopy.

Quality defects such as banding, leaning, and friable tablets are common. SLA and selective laser sintering can introduce variability in layer thickness and composition, consistent fusion, laser-induced degradation, warping, and residue accumulation. Control over raw materials and intermediate properties, such as filament consistency, distribution of polymer molecular weights, the glass transition temperature, and viscoelastic properties significantly reduces manufacturing issues. NIR spectroscopy facilitates real-time monitoring in this regard [32, 71]. Problems with clogs, homogeneity, layer delamination, incorrect melting, and alignment can arise with extrusion and FDM. For consistent manufacturing, precise control over parameters including head speed, pressure, temperature, and nozzle diameter is essential. In essence, ensuring uniform print droplet formation, efficient drying, and meticulous control over raw materials and processes are fundamental for successful 3DP endeavors [72, 73]. Additionally, print cartridges, potentially containing drug-loaded formulations, face regulatory uncertainty in both the US and EU. If the ink within these cartridges is a formulated drug or a binder solution without the drug, it may be classified as an intermediate raw material, subject to standard regulations and good manufacturing practices (GMP). Compliance with standard CMC requirements is essential for the content of print cartridges. These cartridges are integral components of 3DP equipment and can vary in design and operational principles. Therefore, when drug-loaded or excipient-loaded print cartridges are provided at a printing facility, it is crucial for the facility to implement appropriate controls to ensure the safe and effective printing of final dosage forms. Additionally, the facility must establish robust cleaning and maintenance procedures to prevent any cross-contamination between different drug-loaded or excipient-loaded print cartridges.To guarantee the quality and safety of the final dosage forms, the facility must possess a comprehensive understanding of the materials used in the print cartridges and their compatibility with the printing process. This knowledge is vital for maintaining regulatory compliance and ensuring the integrity of the printed products. Without this understanding, the facility may risk producing substandard or contaminated products [74, 75].

7.4.3 CONSIDERATION FOR 3D PRINTING QUALITY THROUGH REAL-TIME PROCESS MONITORING AND PAT

More specifically, real-time monitoring on an industrial scale during 3DP has the potential to greatly improve productivity and guarantee constant product quality. Every layer deposited during the continuous AM process of 3DP adds to the quality and requirements of the finished product. The absence of real-time monitoring could result in delayed corrective actions. PAT can be applied, albeit on a smaller scale, for on-demand manufacturing or during the pilot scale development of drug products, aiding in understanding process design and linking critical quality attributes of the product. Through PAT implementation, manufacturers can gain insights into process parameters directly impacting product quality, enabling real-time adjustments to ensure consistency and efficiency. This proactive approach can lead to cost reductions, waste minimization, and ultimately, heightened customer satisfaction [76, 77].

Both conventional techniques and 3DP are acceptable for the manufacture of pharmaceutical medicinal goods according to the FDA's industry PAT guidelines. PAT facilitates process comprehension and the establishment of suitable control strategies, offering real-time quality assessment comparable to or surpassing laboratory-based testing on sampled specimens. Spectrophotometers, ultrasonic instruments, and lasers are some of the PAT instruments that can be used in 3DP. These tools provide invaluable insights into the printing process, enabling real-time adjustments to uphold product quality and consistency. By leveraging PAT in 3DP, manufacturers can optimize processes, reduce waste, and enhance efficiency. Furthermore, without the need for conventional analytical testing, mathematical models utilizing PAT may be created for the real-time monitoring of quality parameters including assay, dissolution, contaminants, and batch release.

This simplifies manufacturing and saves time and money. In summary, the integration of PAT in 3DP has the potential to revolutionize the industry by elevating quality control and operational efficiency [32, 78].

7.4.4 QUALITY CONTROL AND MATERIAL MANAGEMENT IN 3D PRINTING PHARMACEUTICAL MANUFACTURING

3DP has emerged as a novel platform for pharmaceutical bulk production, initially embraced for personalized medicine on-demand. The FDA's approval of Spritam marks a significant milestone in its adaption for commercial bulk manufacturing. Despite its innovative nature, the 3DP process is expected to adhere to stringent controls akin to traditional pharmaceutical manufacturing. However, instances may arise where products are removed due to software glitches, printer misalignments, or entire batch rejection. Establishing clear criteria for product acceptance, rejection, and batch disposal is imperative to ensure desired quality and consistency [79, 80].

Material carryover presents a common challenge in 3DP, with the quantity of carried-over material influenced by the printer type. For instance, in powder IJ 3DP, over half of the material may persist after the build cycle. Unbound materials are reclaimed and reintroduced into the build cycle. It is vital to thoroughly characterize these carried-over materials for purity, impurities, degradation, uniformity, flowability, moisture content, and other relevant physicochemical properties specific to the 3DP process [81].

7.4.5 SETTING SPECIFICATIONS AND SCALING UP IN 3D PRINTED DRUG MANUFACTURING

The ICH Q6 guideline should be followed when establishing specifications for 3D printed pharmaceutical items. This guideline targets routine bulk pharmaceutical production, requiring a clinically relevant specification for 3DP of standard dosage forms like capsules, tablets, parenteral patches, ointments, and suspensions, as described in compendial pharmacopoeias (USP, BP, and JP), to ensure all essential quality attributes are met. Additionally, specific quality attributes unique to products developed via 3DP, such as tablet porosity impacting tensile strength, may need to be considered. Conducting simulated shipping studies on these 3D printed dosage forms can help ensure product integrity during distribution, shipping, and storage. Stability studies should also be carried out to evaluate the shelf-life of these 3D printed dosage forms in diverse environmental settings. It's crucial to take into account how packing materials affect the product's overall stability and quality. A similar process should be followed when defining requirements for customized or on-demand production in situations such as drug shortages, epidemics, or wars. On the other hand, several characteristics listed in the specification could change depending on the patient or the circumstance. In severe circumstances such as conflict, when the immediate need for life-saving pharmaceutical treatments takes precedence, some qualities may be judged redundant, and personalized dosage must be customized to particular patients or patient groups. In these cases, flexibility in the manufacturing process is crucial to quickly adapt to changing needs and ensure timely delivery of essential medications. It is important to prioritize critical quality attributes while still maintaining efficiency and speed in production.

An industrial-sized IJ 3D printer is one example of larger gear with more print heads that has a substantial impact on the finished product's functioning and quality. Scale-up problems can be impacted by changes in nozzle diameter, pressure, temperature, head speed, material flow rate, cooling rate, build speed, and energy density. Individual layer quality requirements can be impacted by these difficulties, which can be efficiently controlled by real-time monitoring of physicochemical parameters. Utilizing advanced monitoring technologies can help ensure consistent quality and performance throughout the scale-up process. It is important to continuously optimize and adjust parameters to meet desired outcomes and maintain efficiency in large-scale AM operations [82, 83].

7.4.6 POST-PROCESSING AND PACKAGING FOR 3D PRINTED PHARMACEUTICALS

Post-processing and packaging in 3DP encompass a range of operations that directly influence the final product quality. Each type of 3D printer may entail distinct post-processing unit operations, including drying, powder material harvesting, recycling, and packaging. Even if the particular kind of 3D printer being used upstream may not have a direct bearing on these processes, resolving possible problems necessitates a detailed knowledge of the underlying reason.

It is crucial to recognize every potential negative impact of post-processing and put the right mitigation techniques in place in order to reduce any potential negative impacts. The 3DP process's operation qualification should include unique parameters for control limits, correlated with critical quality attributes through a systematic QbD approach, enhancing understanding of process risks and failure mechanisms [84, 85].

Handling non-conforming materials resulting from deviations is integral to pharmaceutical quality systems. Established procedures for detecting and removing non-conforming material must be in place. The process control and monitoring system for 3DP has to be built to recognize typical operations and reroute non-conforming material appropriately. In summary, post-processing and packaging are critical components of pharmaceutical quality systems in 3DP. By adhering to these principles, manufacturers can ensure the safety and quality of their products [63].

7.4.7 MEDICAL PRODUCTS REGULATION

Medical items, such as drugs and equipment, are protected by legal frameworks that are specific to different countries. These frameworks serve as guidelines for manufacturers, ensuring the safe production and distribution of medical products while minimizing risks to patients. Incorporating 3DP into the health and medical sectors necessitates compliance with these frameworks, provided that the printed items or the printers themselves qualify as medical products [86].

Although there is some utilization of 3DP in medicine, particularly in fabricating medications, it remains in its early and experimental phases. Nevertheless, the potential of 3DP in pharmaceuticals holds promise for revolutionizing the manufacturing process by offering a straightforward and rapid method for producing customized small-scale or unique batches. Another experimental application of 3DP lies in bioprinting, which could fall under various regulatory categories such as drugs, biologics, medical devices, or combination products. Interestingly, the definition of "medical devices" in the context of MDR is already widely used in 3DP settings; such hobbyist-focused websites like Thingiverse, where design files for a range of medical devices are easily accessible. Legally speaking, there isn't a single, worldwide statute that regulates medical devices. Rather, technological standards are crucial, especially when it comes to device testing in the 3DP industry for health and medicine. On the other hand, although not without controversy, the TRIPS Agreement establishes worldwide legal norms in the field of intellectual property (IP). To navigate medical regulation, one must refer to specific national and regional regulatory frameworks, such as those overseen by the FDA in the US and the harmonized regulations within the European Union. Generally, MDR follows a risk-based strategy, meaning that devices that pose a greater danger are subject to stricter regulations. The COVID-19 pandemic has had a major impact on the usage of 3DP in healthcare and medical items, as evidenced by the US FDA's instructions on 3DP regulations for custom-made and patient-matched devices. Growing numbers of makerspaces and people with 3D printers producing personal protective equipment (PPE) highlight the necessity of better knowledge and awareness of regulatory frameworks, particularly in non-traditional manufacturing contexts [87].

Troxler's (2022) study on the maker community's response to COVID-19 highlighted a preference for collaboration with medical professionals and hospitals to ensure compliance with regulatory requirements. However, meaningful engagement with regulatory bodies remains a challenge for makers. Moving forward, medical product regulators should proactively engage with the maker community to ensure regulatory compliance and gather feedback from this non-traditional sector of manufacturers [88].

7.5 REGULATORY CHALLENGES IN PERSONALIZED MEDICINE MANUFACTURING USING 3D PRINTING TECHNOLOGY

The manufacturing of personalized medicine via 3DP technology is encountering significant regulatory gaps and hurdles. Although the present regulatory frameworks facilitate industrial-scale 3DP manufacturing, the main obstacle is putting personalized prescription medicinal goods into non-traditional locations including homes, pharmacies, hospitals, and doctors' offices. This calls for a thorough evaluation of the regulatory frameworks now in place in all nations, but especially in the West, that govern the production, distribution, and compounding of pharmaceuticals management in primary care: lessons for Australia from the international experience. *The Medical Journal of Australia*, *216*(9), 445–448. https://doi.org/10.5694/mja2.51497.

3. Nagarajan, N., Dupret-Bories, A., Karabulut, E., Zorlutuna, P., & Vrana, N. E. (2018). Enabling personalized implant and controllable biosystem development through 3D printing. *Biotechnology Advances, 36*(2), 521–533. https://doi.org/10.1016/j.biotechadv.2018.02.004.
4. Vignali, V., Hines, P. A., Cruz, A. G., Ziętek, B., & Herold, R. (2022). Health horizons: Future trends and technologies from the European Medicines Agency's horizon scanning collaborations. *Frontiers in Medicine (Lausanne)*, *9*, 1064003. https://doi.org/10.3389/fmed.2022.1064003.
5. Elton, J., & O'Riordan A. (2016). *Healthcare disrupted: Next generation business models and strategies.* Hoboken, New Jersey: John Wiley & Sons, Inc.
6. Zagalo, D. M., Sousa, J., & Simões, S. (2022). Quality by design (QbD) approach in marketing authorization procedures of Non-biological complex drugs: A critical evaluation. *European Journal of Pharmaceutics and Biopharmaceutics*, 178 1–24. https://doi.org/10.1016/j.ejpb.2022.07.014.
7. Adalbert, L., Kanti, S. P. Y., Jójárt-Laczkovich, O., Akel, H., & Csóka, I. (2022). Expanding quality by design principles to support 3D Printed medical device development following the renewed regulatory framework in Europe. *Biomedicines, 10*(11), 2947. https://doi.org/10.3390/biomedicines10112947.
8. Zahir, A., Mahmood, U., Nazir, A., Hussain, T., & Abid, S. (2022). Biomaterials for medical and healthcare products. In *Medical textiles from natural resources: The textile institute book series* (pp. 43–86). Woodhead Publishing. https://doi.org/10.1016/B978-0-323-90479-7.00013-0.
9. Leary, M. (2020). Binder jetting. In *Design for additive manufacturing: Additive manufacturing materials and technologies* (pp. 335–339). Elsevier. https://doi.org/10.1016/b978-0-12-816721-2.00013-0.
10. Miyanaji, H. (2018). *Binder jetting additive manufacturing process fundamentals and the resultant influences on part quality* (Ph.D. Dissertation, Louisville, Kentucky).
11. Blunk, H., & Seibel, A. (2023). Design guidelines for metal binder jetting. *Progress in Additive Manufacturing*, *9*, 725–732. https://doi.org/10.1007/s40964-023-00475-y.
12. Gibson, I., Rosen, D., & Stucker, B. (2015). Directed energy deposition processes. In: *Additive manufacturing technologies* (pp. 245–268). New York, NY: Springer. https://doi.org/10.1007/978-1-4939-2113-3_10.
13. Ahn, D. G. (2021). Directed Energy Deposition (DED) Process: State of the Art. *International Journal of Precision Engineering and Manufacturing - Green Technology*, *8*, 703–742. https://doi.org/10.1007/s40684-020-00302-7.
14. Saboori, A., Aversa, A., Marchese, G., Biamino, S., Lombardi, M., & Fino, P. (2019). Application of directed energy deposition-based additive manufacturing in repair. *Applied Sciences*, *9*(16), 3316. https://doi.org/10.3390/app9163316.
15. Loh, G. H., Pei, E., Gonzalez-Gutierrez, J., & Monzón, M. (2020). An overview of material extrusion troubleshooting. *Applied Sciences (Switzerland)*, *10*(14). https://doi.org/10.3390/app10144776.

16. Haghighi, A. (2023). Material extrusion. In E. Pei, *et al* (Eds)., *Springer handbook of additive manufacturing* (pp. 335–347). Cham: Springer Handbooks, Springer. https://doi.org/10.1007/978-3-031-20752-5_21.
17. Leary, M. (2020). Material jetting. In *Design for additive manufacturing* (pp. 269–282). https://doi.org/10.1016/b978-0-12-816721-2.00009-9.
18. Kroczek, K., Turek, P., Mazur, D., Szczygielski, J., Filip, D., Brodowski, R., Balawender, K., Przeszlowski, Ł., Lewandowski, B., Orkisz, S., Mazur, A., Budzik, G., Cebulski, J., & Oleksy, M. (2022). Characterisation of selected materials in medical applications. *Polymers (Basel)*, *14*(8), 1526. https://doi.org/10.3390/polym14081526.
19. Behera, A. (2022). Additive manufacturing materials. In *Advanced materials: An introduction to modern materials science* (pp. 667–700). Cham: Springer International Publishing. https://doi.org/10.1007/978-3-030-80359-9_20.
20. Popov, V. V., Grilli, M. L., Koptyug, A., Jaworska, L., Katz-Demyanetz, A., Klobčar, D., Balos, S., Postolnyi, B. O., & Goel, S. (2021). Powder bed fusion additive manufacturing using critical raw materials: A review. *Materials (Basel)*, *14*(4), 1–37. https://doi.org/10.3390/ma14040909.
21. Datta, B. K. (2011). *Powder metallurgy: An advanced technique of processing engineering materials* (pp. 204). PHI Learning Pvt. Ltd. http://books.google.com/books?id=Sr3gldoqQIoC&pgis=1.
22. Ligon, S. C., Liska, R., Stampfl, J., Gurr, M., & Mülhaupt, R. (2017). Polymers for 3D printing and customized additive manufacturing. *Chemical Reviews*, *117*(15), 10212–10290.
23. Mahamood, R. M., & Akinlabi, E. T. (2017). Additive manufacturing of funtionally graded materials, top. mining. *Metallurgical and Materials Engineering*, 47–68. https://doi.org/10.1007/978-3-319-53756-6_4.
24. Salifu, S., Ogunbiyi, O., & Olubambi, P. A. (2022). Potentials and challenges of additive manufacturing techniques in the fabrication of polymer composites. *The International Journal of Advanced Manufacturing Technology*, *122*, 577–600. https://doi.org/10.1007/s00170-022-09976-6.
25. Kharat, V. J., Singh, P., Sharath Raju, G., Kumar Yadav, D., Satyanarayana Gupta, M., Arun, V., Hussein Majeed, A., & Singh, N. (2023). Additive manufacturing (3D printing): A review of materials, methods, applications and challenges. *Materials Today: Proceedings*. In press. https://doi.org/10.1016/j.matpr.2023.11.033.
26. Zhang, F., Zhu, L., Li, Z., Wang, S., Shi, J., Tang, W., Li, N., & Yang, J. (2021). The recent development of vat photopolymerization: A review. *Additive Manufacturing*, *48*, 102423. https://doi.org/10.1016/j.addma.2021.102423.
27. Sui, Y., Radwan, A. N., Gopalakrishnan, A., Dikshit, K., Bruns, C. J., Zorman, C. A., & Whiting, G. L. (2023). A reactive inkjet printing process for fabricating biodegradable conductive zinc structures. *Advanced Engineering Materials*, *25*, 2200529. https://doi.org/10.1002/adem.202200529.
28. Pagac, M., Hajnys, J., Ma, Q. P., Jancar, L., Jansa, J., Stefek, P., & Mesicek, J. (2021). A review of vat photopolymerization technology: Materials, applications, challenges, and future trends of 3D printing. *Polymers (Basel)*, *13*(4), 598. https://doi.org/10.3390/polym13040598.

TABLE 8.1 Polymer and Their Impurity Limit.

Polymer	Allowable Impurity Limit (%)
Polyethylene	≤ 0.1
Polylactic acid	≤ 0.05

In this example, both polyethylene and PLA are commonly used polymers in pharmaceutical FDM. However, the allowable impurity limits differ between the two polymers. For polyethylene, the allowable impurity limit is set at ≤0.1%, while for PLA, it is stricter, with an allowable impurity limit of ≤0.05%. These limits ensure that the polymers meet the required standards for chemical purity, minimizing the risk of impurities affecting the safety and efficacy of the final drug product [6, 7].

Mechanical Properties

The mechanical properties of the materials used in FDM are crucial determinants of the final product's performance. These properties include tensile strength, elasticity, and degradation rate, which need to be thoroughly evaluated through standardized testing procedures. The mechanical integrity of the printed object must be sufficient to withstand physiological conditions while maintaining the required drug release profile. Material testing involves a variety of stress tests and simulations that mimic the conditions the product will face in a real-world biological environment [5, 8].

TABLE 8.2 Advantages and Disadvantages of Drug Compatibility.

Advantages of Drug Compatibility	Disadvantages of Drug Compatibility
1. Enhanced safety: Ensures that drugs can be safely administered together without adverse reactions or interactions.	1. Limited options: May restrict the choice of drugs that can be co-administered, potentially reducing treatment options.
2. Improved efficacy: Allows for the simultaneous administration of multiple drugs, leading to synergistic effects and better treatment outcomes.	2. Complexity: Determining drug compatibility can be complex and time-consuming, requiring careful consideration of various factors.
3. Reduced errors: Minimizes the risk of medication errors by preventing inadvertent mixing of incompatible drugs, thus enhancing patient safety.	3. Compatibility issues: Some drugs may be inherently incompatible with each other due to chemical or physical properties, limiting their simultaneous use.
4. Cost-effective: Facilitates the use of combination therapies, potentially reducing the overall cost of treatment by eliminating the need for separate administrations.	4. Storage challenges: Compatibility requirements may necessitate separate storage conditions for different drugs, increasing logistical complexities.

TABLE 8.2 *(Continued)*

Advantages of Drug Compatibility	Disadvantages of Drug Compatibility
5. Streamlined administration: Simplifies the medication administration process by allowing multiple drugs to be administered together, improving patient compliance.	5. Monitoring requirements: Requires diligent monitoring of patients for any signs of adverse reactions or interactions when administering multiple drugs simultaneously.

Drug Compatibility

One of the most critical aspects of material selection in pharmaceutical FDM is ensuring compatibility between the drug and the polymer matrix. The interaction between the drug and the materials should not result in any adverse chemical reactions or degradation of the drug during the printing process. Stability testing, including thermal and photostability studies, are vital to determine if the high temperatures used in FDM affect the drug's potency or safety profile.

In addition, the potential for any chemical interaction that might occur needs to be evaluated. This includes studying the potential for hydrolysis, oxidation, or any other degradation pathways that could be catalyzed by the presence of the polymer. Moreover, the solubility of the drug in the polymer, its diffusion rate, and the eventual release mechanism are critical parameters that influence the effectiveness of the dosage form. These aspects are usually assessed through *in vitro* dissolution testing followed by appropriate *in vivo* studies to confirm the modeling predictions [6–9].

TABLE 8.3 Drugs and Their Compatibility.

Drug 1	Drug 2	Compatibility
Aspirin	Ibuprofen	Incompatible
Amoxicillin	Clavulanic acid	Compatible
Amlodipine	Lisinopril	Compatible
Warfarin	Clopidogrel	Potentially compatible
Simvastatin	Gemfibrozil	Incompatible

In this example, each row represents a pair of drugs, and the "Compatibility" column indicates whether they are compatible or incompatible with each other. For instance, aspirin and ibuprofen are marked as incompatible, while amoxicillin and clavulanic acid are compatible. The compatibility information helps healthcare professionals make informed decisions about medication administration to minimize the risk of adverse reactions or interactions [10–12].

8.2.2 REGULATORY COMPLIANCE AND DOCUMENTATION

Given the direct impact of these materials on patient health, regulatory compliance cannot be overstressed. Documentation detailing the manufacturing process, material sourcing, and compliance with pharmacopeial standards is essential for regulatory submissions. The FDA requires detailed information on the type of material, its source, batch consistency, and the specific conditions used during the manufacturing process [14, 15].

Materials used in pharmaceutical FDM must also adhere to the principles of current GMPs (cGMP) to ensure that products are consistently produced and controlled according to quality standards. This encompasses every aspect of the manufacturing process, from the initial material selection and handling through to the final product testing and release [23, 24].

8.3 FDM PROCESS PARAMETERS

The production of pharmaceuticals using FDM technology introduces a layer of complexity with respect to ensuring that both the physical and chemical properties of the printed objects meet stringent pharmaceutical standards. The process parameters of FDM printers must be meticulously controlled to fabricate dosage forms that are not only pharmaceutically efficacious but also meet critical quality attributes. This segment focuses on the critical parameters that influence the quality of pharmaceutical products and the in-process controls necessary to maintain the integrity of the manufacturing process [14–16].

TABLE 8.4 Different Process Parameter Used in FDM 3D Printing.

Process Parameter	Description	Range/Value
Nozzle temperature	Temperature of the extrusion nozzle	180–260°C
Build plate temperature	Temperature of the build plate	20–110°C
Layer height	Height of each printed layer	0.1–0.3 mm
Printing speed	Speed at which the extruder moves	20–100 mm/s
Infill density	Percentage of infill material in the print	10–80%
Filament diameter	Diameter of the filament used	1.75 or 2.85 mm

The "Description" column briefly describes what each parameter entails, while the "Range/Value" column provides either a range or a specific value for each parameter. These parameters play crucial roles in determining the quality, strength, and dimensional accuracy of the printed parts in FDM [16–19].

8.3.1 CRITICAL PARAMETERS

8.3.1.1 EXTRUSION TEMPERATURE

The extrusion temperature in FDM is a pivotal parameter, as it needs to be precisely managed to ensure the polymer melts properly without degrading any APIs incorporated into the filament. An optimal temperature setting is crucial because excessive heat can lead to the decomposition of sensitive APIs, altering their effectiveness or safety. Conversely, temperatures that are too low may result in improper melting of the polymer, leading to poor layer adhesion and compromised structural integrity of the final product. Calibration of the extrusion temperature is typically conducted through initial trials followed by adjustments based on the stability outcomes of the APIs and the mechanical properties of the extrudate [23, 24].

8.3.1.2 PRINT SPEED AND PATTERN

The speed at which the printer head moves and the pattern it follows during deposition significantly affect the porosity and mechanical strength of the final product. These factors, in turn, influence the drug release rates from the printed forms. A faster print speed can lead to insufficient time for the proper adhesion of layers, thereby affecting the print's density and uniformity. In contrast, slower speeds may increase the density, potentially reducing porosity to undesirable levels which could retard drug release. The print pattern, or the path followed by the nozzle, can be adjusted to modify the microstructure of the print, tailoring the porosity and thus controlling the release kinetics of the API [19–21].

8.3.1.3 LAYER THICKNESS

Layer thickness determines the resolution of the printed object, impacting both the surface texture and the structural integrity. Thinner layers typically result in smoother surfaces and more detailed prints, which can enhance the dissolution profile of the drug product. However, reducing layer thickness might increase the print time and affect the mechanical strength of the object. Optimizing layer thickness is thus a balance between achieving desired surface characteristics and maintaining production efficiency and object robustness.

8.3.2 IN-PROCESS CONTROLS

8.3.2.1 REAL-TIME MONITORING

To ensure consistent quality and process stability, real-time monitoring technologies are employed. Techniques such as thermal imaging can be used to monitor the temperature of the extrusion head and the deposited material, ensuring that it remains within a range that is safe for the APIs. Acoustic monitoring can detect changes in the sound emitted by the printer, which may indicate mechanical issues or deviations in the print material properties that could affect product quality. These monitoring tools help in detecting anomalies early in the printing process, allowing for timely interventions.

8.3.2.2 POSTPRINT INSPECTION

After printing, rigorous inspections are performed to validate the quality of the product. Dimensional checks ensure that the printed object meets the specified geometrical criteria. Disintegration testing under simulated physiological conditions is crucial for confirming that the product will behave as expected *in vivo*. Mechanical strength assessments are also conducted to ensure that the object can withstand handling and does not break down prematurely or inconsistently, which could lead to dosage errors [15, 17].

8.4 VALIDATION AND QUALITY ASSURANCE

Process validation is a critical component in pharmaceutical FDM, structured around three core phases: Installation Qualification (IQ), Operational Qualification (OQ), and Performance Qualification (PQ).

8.4.1 INSTALLATION QUALIFICATION (IQ)

The first step, IQ, involves a thorough check to ensure that the FDM equipment has been delivered, installed, and set up according to the manufacturer's specifications. This stage verifies that the physical setup complies with design specifications and that all supporting systems (such as software or environmental controls) are installed and configured correctly. Documentation from this phase includes reports on equipment delivery, installation, configuration, and calibration records.

8.4.2 OPERATIONAL QUALIFICATION (OQ)

OQ is the subsequent phase where the equipment is tested to confirm it operates within the defined parameters for expected performance. OQ tests the machine's ability to function according to its operational specifications under various conditions. This includes verifying that the printer maintains the correct extrusion temperature, print speed, and pattern accuracy over a range of operating conditions. The documentation for OQ includes test results for all operational limits and conditions, including failure modes and any corrective actions taken.

8.4.3 PERFORMANCE QUALIFICATION (PQ)

The final phase, PQ, involves validating that the FDM process consistently produces acceptable products under normal operation conditions. PQ is typically conducted under real production conditions with actual product formulations. It includes extensive sampling and testing of the final products to ensure they meet all release criteria, are reproducible, and adhere to predetermined quality attributes. Documentation from PQ encompasses batch records, quality control results, and statistical analysis demonstrating process consistency and control [20, 24].

8.5 QUALITY ASSURANCE PROTOCOLS

Quality assurance in pharmaceutical FDM extends beyond validation to include continuous monitoring and control of the production process.

8.5.1 BATCH DOCUMENTATION

Maintaining detailed records for each production batch is essential for traceability and quality control. Batch documentation includes the recording of material lot numbers, specific machine settings used, environmental conditions during production, and results from in-process and post-process inspections. This comprehensive documentation ensures that each step of the production process is reproducible and traceable, and facilitates the investigation and rectification of deviations from quality standards.

8.5.2 STATISTICAL PROCESS CONTROL (SPC)

Statistical process control (SPC) is employed to monitor and control the manufacturing process using statistical methods. SPC can identify process trends and variations in real-time, which enables timely adjustments to maintain process control and product quality. By analyzing data collected from the process, manufacturers can detect potential problems early, thereby preventing defects and ensuring that the process remains within specified control limits. This proactive approach not only enhances product quality but also reduces production costs by minimizing waste and rework [19–22].

8.5.3 REGULATORY CONSIDERATIONS

The regulatory landscape for pharmaceutical FDM is intricate and comprehensive, involving stringent guidelines to ensure product safety, efficacy, and quality. Regulatory bodies such as the U.S. FDA and the European Medicines Agency (EMA) provide the framework under which pharmaceutical products manufactured using FDM must operate. These regulations cover a broad spectrum of considerations, including the safety of materials used, the efficacy of the pharmaceutical products, the environmental impact of the manufacturing processes, and the safety of operators [24, 25].

8.6 COMPLIANCE AND STANDARDIZATION

Adherence to international standards and frameworks is crucial in meeting regulatory requirements and ensuring the reliability of pharmaceutical products produced via FDM.

8.6.1 ISO STANDARDS

ISO 13485 is a globally recognized standard that specifies requirements for a quality management system (QMS) specific to the medical device industry, which encompasses pharmaceutical FDM systems. Adhering to ISO 13485 helps organizations demonstrate their ability to provide medical devices and related services that consistently meet customer and applicable regulatory requirements. This standard focuses on risk management and risk-based decision-making processes, which are critical when implementing new manufacturing technologies like FDM. It includes provisions for the design, development, production, and delivery of medical devices, ensuring comprehensive quality oversight throughout the product lifecycle.

Implementing ISO 13485 provides a structured framework to ensure product quality and safety, helping manufacturers meet the stringent requirements of regulatory authorities. It assists in defining clear procedures for all stages of manufacturing, from material selection and machine calibration to final product testing and traceability [24, 26].

8.6.2 RISK MANAGEMENT

ISO 14971 provides a thorough framework for manufacturers to identify, evaluate, control, and monitor the risks associated with medical devices, including those produced by FDM. Given the novel nature of pharmaceutical FDM, identifying potential risks from material inputs, software algorithms, equipment malfunctions, or operator errors is essential. ISO 14971 helps manufacturers implement a systematic approach to risk management, which includes the assessment of potential hazards associated with the FDM process and the formulation of strategies to mitigate these risks.

This risk management process is continuous and involves several steps:

1. Risk Analysis: Identify hazards and hazardous situations associated with the FDM process that could lead to a medical device failure or harm to patients.
2. Risk Evaluation: Assess the potential impact and likelihood of each risk, determining which risks need to be addressed.

3. Risk Control: Implement appropriate measures to minimize or eliminate risks. This could involve changes to the design, manufacturing process, quality controls, or introducing additional safety measures.
4. Evaluation of Overall Residual Risk: Assess whether the overall risk level after implementing controls is acceptable.
5. Risk Management Report: Document the risk analysis, evaluation, control processes, and results in a comprehensive report.
6. Production and Postproduction Information: Continuously monitor and update risk assessments based on feedback and new information obtained from the postproduction phase.

Navigating the regulatory environment for pharmaceutical FDM involves a deep understanding of and compliance with existing standards and guidelines. By adhering to standards like ISO 13485 and ISO 14971, manufacturers can ensure that they not only meet regulatory expectations but also safeguard patient health and safety. Effective implementation of quality management and risk management frameworks is essential for the successful adoption and scaling of FDM technologies in the pharmaceutical industry. This rigorous approach to regulation and compliance will be vital as the technology evolves and becomes more integrated into mainstream pharmaceutical manufacturing processes [24–26].

8.7 FDM 3D PRINTING OF COMMERCIAL FILAMENTS AND DRUG-LOADED FILAMENTS

FDM 3D printing has emerged as a transformative technology, revolutionizing manufacturing processes across various industries. This innovative additive manufacturing technique involves the extrusion of thermoplastic filaments layer by layer, enabling the creation of complex geometries with high precision and customization. In recent years, FDM 3D printing has gained significant traction not only in the production of commercial filaments but also in the development of drug-loaded filaments, offering novel solutions in the fields of pharmaceuticals and healthcare [2–5].

FDM 3D printing operates on the principle of layer-by-layer deposition of thermoplastic materials, driven by computer-aided design (CAD) models. The process begins with the conversion of digital designs into slicing software, which generates a series of cross-sectional layers.

These layers are then translated into machine instructions that control the movement of the print head and the extrusion of molten filament material onto a build platform. As each layer solidifies upon deposition, subsequent layers are added to gradually build the desired object. This layering approach offers unparalleled flexibility in design complexity and customization, making FDM 3D printing suitable for a wide range of applications.

One of the primary applications of FDM 3D printing is the production of commercial filaments for various industries, including automotive, aerospace, consumer goods, and electronics. Manufacturers leverage FDM technology to create prototypes, functional parts, and end-use products with reduced lead times and production costs. Commercial filaments for FDM printing are available in a diverse range of thermoplastic materials, such as acrylonitrile butadiene styrene (ABS), PLA, polyethylene terephthalate glycol (PETG), and nylon, each offering unique properties such as strength, flexibility, and heat resistance. These materials undergo rigorous testing and quality control processes to ensure consistency and reliability in 3D printing applications, meeting the demands of various industrial and consumer needs.

In recent years, there has been a growing interest in leveraging FDM 3D printing technology for the fabrication of drug-loaded filaments, opening new frontiers in pharmaceutical research and personalized medicine. Drug-loaded filaments are filament-based drug delivery systems that incorporate APIs within a biocompatible polymer matrix. The integration of FDM 3D printing enables precise control over drug dosage, release kinetics, and spatial distribution, offering significant advantages over conventional dosage forms such as tablets and capsules. Furthermore, the ability to tailor the physical and mechanical properties of drug-loaded filaments allows for personalized therapeutic solutions and enhanced patient compliance.

The fabrication of drug-loaded filaments involves several key steps, starting from the selection of suitable polymers and APIs based on their compatibility, solubility, and stability. Polymeric carriers such as polyvinyl alcohol (PVA), PCL, and poly (lactic-co-glycolic acid) (PLGA) are commonly used due to their biodegradability and biocompatibility. The API is then dispersed or dissolved within the polymer matrix using various techniques, including physical mixing, solvent casting, or hot-melt extrusion. The resulting drug-loaded filament undergoes characterization tests to evaluate its mechanical properties, drug content, and release profiles, ensuring consistency and efficacy [14, 22].

The development of drug-loaded filaments via FDM 3D printing presents numerous opportunities in the pharmaceutical industry, including personalized drug delivery, on-demand manufacturing, and patient-specific dosage forms. By incorporating multiple APIs or adjusting drug concentrations within the filament, complex therapeutic regimens can be simplified into single-dose formulations, enhancing treatment adherence and efficacy. Moreover, FDM 3D printing enables the fabrication of patient-specific dosage forms tailored to individual anatomical and physiological characteristics, facilitating personalized medicine and optimized therapeutic outcomes [22].

In addition to pharmaceutical applications, drug-loaded filaments produced via FDM 3D printing hold promise in other fields such as tissue engineering, regenerative medicine, and medical device fabrication. These multifunctional constructs can deliver bioactive agents, growth factors, or cells to targeted sites for tissue regeneration, wound healing, and organ repair. By precisely controlling the spatial distribution of therapeutic components within the filament, complex tissue architectures can be recreated with high fidelity, mimicking native tissue structures and functions. Furthermore, the integration of imaging modalities such as computed tomography (CT) and magnetic resonance imaging (MRI) enables the development of patient-specific implants and scaffolds tailored to individual patient anatomy [13–16].

Despite the numerous advantages offered by FDM 3D printing of commercial filaments and drug-loaded filaments, several challenges remain to be addressed to realize its full potential. These include optimizing printing parameters for consistent quality and mechanical properties, ensuring regulatory compliance and safety standards for medical applications, and addressing scalability and cost-effectiveness issues for large-scale production. Additionally, further research is needed to explore advanced materials, multifunctional formulations, and novel printing techniques to enhance the performance and versatility of FDM 3D printing in diverse applications.

In conclusion, FDM 3D printing has emerged as a transformative technology with far-reaching implications in commercial filament production and drug-loaded filament fabrication. By harnessing the power of additive manufacturing, manufacturers can achieve unprecedented levels of design flexibility, customization, and efficiency, revolutionizing traditional manufacturing processes. Moreover, the integration of FDM 3D printing in pharmaceutical research opens new avenues for personalized medicine, targeted drug delivery, and regenerative therapies, offering promising solutions to complex healthcare challenges. As research and innovation continue to advance, FDM 3D printing is poised to reshape industries and improve lives around the globe.

8.7.1 LEVELING OF THE BUILD PLATE OF THE PRINTER

Leveling the build plate of a printer is a critical task in the realm of 3D printing, where precision and accuracy are paramount. Ensuring that the build plate is perfectly aligned not only facilitates successful prints but also contributes to the longevity of the printer and the quality of the printed objects.

At the heart of 3D printing lies the build plate, a foundational component that serves as the canvas upon which intricate creations materialize. The build plate, often made of glass, aluminum, or other flat materials, provides the stable foundation necessary for the deposition of successive layers of material to form a three-dimensional object. However, for this process to unfold seamlessly, the build plate must be perfectly level, ensuring uniform adhesion and layer deposition throughout the printing process [20].

The significance of leveling the build plate cannot be overstated, as even minor deviations can have cascading effects on print quality. Uneven leveling may result in adhesion issues, warped prints, or misalignment of layers, leading to structural weaknesses or even outright failure of the print. Moreover, a misaligned build plate can strain the printer's components, potentially causing premature wear and tear or malfunctions.

To achieve optimal print outcomes, several methods for leveling the build plate have been developed, each with its own merits and intricacies. One common approach involves manual leveling, wherein the user adjusts the height of the build plate at various points until it is parallel to the printer's nozzle. This method typically requires the use of a piece of paper or a feeler gauge to gauge the distance between the nozzle and the build plate, adjusting the leveling screws accordingly.

Another method gaining popularity is automatic bed leveling, wherein sensors integrated into the printer probe the build plate's surface to map out any deviations. Through this process, the printer compensates for unevenness by dynamically adjusting the print bed's position during the printing process. Automatic bed leveling offers convenience and accuracy, particularly for novice users or those dealing with intricate print geometries [21, 22].

Despite the advancements in leveling technology, challenges and complications may still arise during the process. One common issue is bed adhesion, where prints fail to adhere uniformly to the build plate, leading to warping or detachment mid-print. This problem can often be mitigated through adjustments to the build plate's temperature, surface treatment, or the use of adhesion aids such as hairspray or adhesive sheets.

Additionally, mechanical issues such as backlash, worn components, or improper assembly can contribute to difficulties in leveling the build plate. Regular maintenance and calibration of the printer's mechanical systems are crucial to pre-empting such issues and ensuring consistent print quality over time.

Looking ahead, ongoing research and development efforts are poised to further streamline the process of leveling the build plate and enhance overall print reliability. Advancements in sensor technology, machine learning algorithms, and material science hold the promise of automated systems capable of detecting and correcting leveling discrepancies in real-time, reducing the burden on users and expanding the accessibility of 3D printing technology [25–27].

In conclusion, leveling the build plate of a printer is a foundational aspect of 3D printing, essential for achieving consistent print quality and reliability. Whether through manual adjustments or automated systems, the meticulous alignment of the build plate sets the stage for successful prints and unlocks the full potential of this transformative technology. As innovation continues to drive progress in the field of 3D printing, the quest for perfection in build plate leveling remains an enduring pursuit, emblematic of humanity's ongoing quest for precision and excellence in craftsmanship.

8.8 CONCLUSION

The integration of FDM into pharmaceutical manufacturing presents a promising avenue for creating more tailored and effective drug delivery systems. However, the successful implementation of this technology hinges critically on adhering to stringent quality control and validation protocols. Ensuring the reproducibility of drug formulations, maintaining the integrity and precision of printed products, and meeting regulatory compliances are essential steps in leveraging FDM's full potential. As the technology advances, continuous improvement of material characterization, process validation, and real-time monitoring will be vital. Collaborative efforts between regulatory bodies, industry experts, and academic researchers will also play a crucial role in establishing standardized guidelines that safeguard patient health while fostering innovation in pharmaceutical manufacturing through FDM. This collaborative, rigorous approach will not only enhance the reliability and safety of FDM-produced pharmaceuticals but also help realize the technology's transformative potential in personalized medicine.

KEYWORDS

- **FDM technology**
- **layer adhesion**
- **pharmaceutical 3D printing**
- **print uniformity**
- **process validation**
- **quality control**

REFERENCES

1. Goyanes, A., Buanz, A., Basit, A. W., & Gaisford, S. (2015). Fused-filament 3D printing (3DP) for fabrication of tablets. *International Journal of Pharmaceutics*, *496*(2), 414–423.
2. Ventola, C. L. (2014). Medical applications for 3D printing: Current and projected uses. *Pharmacy and Therapeutics*, *39*(10), 704–711.
3. Trenfield, S. J., Awad, A., Madla, C. M., Hatton, G. B., Firth, J., & Goyanes, A. (2018). 3D printing pharmaceuticals: Drug development to frontline care. *Trends in Pharmacological Sciences*, *39*(5), 440–451.
4. Sun, J., & Zhou, W. (2020). The application of FDM 3D printing technology in pharmaceutical sciences. *Innovation in Pharmacy*, *11*(1).

CHAPTER 9

Case Studies Demonstrating Pharmaceutical FDM Applications

MRUNAL KARANJKAR[1*], LALITA NEMADE[2], PRATIKSHA NIKAM[2], and ARCHANA HIWASE[3]

[1]*Department of Pharmacology, Govindrao Nikam College of Pharmacy, Sawarde, Maharashtra, India*

[2]*Department of Pharmaceutics, Govindrao Nikam College of Pharmacy, Sawarde, Maharashtra, India*

[3]*Department of Pharmaceutics, Faculty of Pharmacy, B.K. Patil Institute of Pharmacy, Taloja, Navi Mumbai, Maharashtra, India*

*Corresponding author

ABSTRACT

Fused deposition modeling (FDM) has emerged as a transformative technology in pharmaceutical science, offering innovative solutions to critical challenges in drug development, manufacturing, and personalized medicine. This chapter presents a series of real-world case studies demonstrating the diverse applications of FDM in the pharmaceutical industry, including customized drug delivery systems, rapid prototyping, on-demand manufacturing, tissue engineering, and continuous production. These case studies highlight the capacity of FDM to create patient-specific dosage forms with precise drug release kinetics, enhance product development through iterative design, and support agile manufacturing processes with reduced waste and improved efficiency. Furthermore, the integration of FDM in regenerative medicine applications—such as the fabrication of biocompatible scaffolds and implants—underscores its potential in advancing tissue repair and

Advancements in Fused Deposition Modeling (FDM) Technology for Pharmaceutical and Medical Applications. Kuldeep Vinchurkar, Dinesh Kumar Mishra, Pankaj V Dixit, and Sheetal Mane (Eds.)

DOI: 10.1201/9781779640338-9

manufacturing methods may struggle to consistently achieve the stringent standards set by regulatory agencies, leading to compliance issues and quality control challenges.

9.1.7 ROLE OF FDM TECHNOLOGY

FDM technology offers a promising solution to these pharmaceutical challenges by providing a versatile, cost-effective, and scalable manufacturing platform. FDM allows for the precise deposition of materials in layer-by-layer fashion, enabling the fabrication of complex geometries, customized formulations, and controlled-release drug delivery systems [6].

9.2 OBJECTIVE OF CASE STUDIES

The primary objective of case studies demonstrating pharmaceutical FDM applications is to showcase how this innovative technology addresses the aforementioned challenges in the pharmaceutical industry [7]. By presenting real-world examples and empirical evidence, these case studies aim to illustrate the efficacy, feasibility, and potential benefits of integrating FDM into pharmaceutical workflows. Through rigorous experimentation, data analysis, and critical evaluation, these case studies seek to provide actionable insights and practical recommendations for leveraging FDM technology to enhance drug development, manufacturing, and delivery in the pharmaceutical sector [8]. Demonstrating pharmaceutical FDM applications involves showcasing how this additive manufacturing technology is utilized within the pharmaceutical industry to address various challenges and opportunities. Here's an overview of demonstrating pharmaceutical FDM applications [7–9].

9.2.1 CUSTOMIZED DRUG DELIVERY SYSTEMS

Highlight how FDM is used to fabricate personalized drug delivery systems tailored to individual patient needs. Showcase examples of 3D printed dosage forms with precise dosages, release profiles, and geometries optimized for specific therapeutic applications [10].

9.2.2 *RAPID PROTOTYPING AND ITERATIVE DESIGN*

Demonstrate how FDM enables rapid prototyping of pharmaceutical products, such as medical devices, packaging, and equipment components. Showcase how companies iterate and refine designs quickly, accelerating the development process and reducing time-to-market [11].

9.2.3 *ON-DEMAND MANUFACTURING*

Illustrate how FDM facilitates on-demand manufacturing of pharmaceutical products, including spare parts, replacement components, and customized medical devices. Highlight examples of companies leveraging FDM to produce critical items with short lead times, improving operational efficiency and supply chain resilience [12].

9.2.4 *TISSUE ENGINEERING AND REGENERATIVE MEDICINE*

Showcase FDM's role in tissue engineering and regenerative medicine applications within the pharmaceutical industry. Highlight examples of 3D printed scaffolds, implants, and tissue constructs designed to support cell growth, tissue regeneration, and drug testing [13].

9.2.5 *CONTINUOUS MANUFACTURING*

Discuss how FDM-based continuous manufacturing processes are utilized in pharmaceutical production. Showcase examples of continuous fabrication of oral solid dosage forms, transdermal patches, and implants with uniform drug distribution and controlled release profiles [7].

9.2.6 *QUALITY CONTROL AND VALIDATION*

Emphasize the importance of quality control and validation in pharmaceutical FDM applications. Showcase how companies ensure the accuracy, precision, and reproducibility of 3D printed pharmaceutical products through rigorous testing and validation protocols [9, 14].

9.2.7 REGULATORY COMPLIANCE

Highlight the regulatory considerations associated with pharmaceutical FDM applications. Showcase examples of companies navigating regulatory requirements and obtaining approvals for 3D printed drug products, medical devices, and tissue-engineered constructs [14].

9.2.8 COLLABORATIVE PARTNERSHIPS AND INNOVATION

Illustrate how collaborative partnerships between pharmaceutical companies, academic institutions, and technology providers drive innovation in FDM applications. Showcase examples of successful collaborations leading to breakthroughs in drug delivery, tissue engineering, and manufacturing processes [15].

9.3 NEED FOR CASE STUDIES DEMONSTRATING PHARMACEUTICAL FDM APPLICATIONS

The need for case studies demonstrating pharmaceutical FDM applications arises from several factors [3–6, 16].

9.3.1 VALIDATION OF TECHNOLOGY

Case studies provide empirical evidence of the efficacy and feasibility of using FDM in pharmaceutical settings. They validate the technology's potential to address specific challenges and meet industry standards, thereby instilling confidence among stakeholders, including regulators, investors, and decision-makers.

9.3.2 KNOWLEDGE SHARING AND BEST PRACTICES

By documenting successful FDM applications in pharmaceuticals, case studies facilitate knowledge sharing and dissemination of best practices within the industry. They offer valuable insights into implementation strategies, technical considerations, and lessons learned, helping other companies and researchers navigate similar projects more effectively.

9.3.3 BENCHMARKING AND PERFORMANCE EVALUATION

Case studies enable benchmarking of FDM technologies and processes against established benchmarks and industry standards. They allow companies to assess the performance of their FDM systems in terms of quality, efficiency, and cost-effectiveness, facilitating continuous improvement and optimization.

9.3.4 ADDRESSING REGULATORY REQUIREMENTS

Regulatory bodies often require empirical evidence and real-world examples to assess the safety, efficacy, and quality of new technologies and processes introduced in pharmaceutical manufacturing. Case studies demonstrating successful FDM applications help companies fulfill regulatory requirements and obtain necessary approvals for commercialization.

9.3.5 MARKET DIFFERENTIATION AND COMPETITIVE ADVANTAGE

Companies that pioneer and showcase innovative FDM applications in pharmaceuticals can gain a competitive edge in the market. Case studies serve as compelling marketing tools, highlighting the company's expertise, leadership, and commitment to innovation, thereby attracting customers, partners, and investment opportunities.

9.3.6 IDENTIFICATION OF UNTAPPED OPPORTUNITIES

Through case studies, companies and researchers can identify untapped opportunities and unmet needs in the pharmaceutical sector where FDM could offer transformative solutions. By analyzing successful case studies and industry trends, stakeholders can prioritize research and development efforts and allocate resources effectively.

In summary, case studies demonstrating pharmaceutical FDM applications play a crucial role in validating technology, sharing knowledge, benchmarking performance, addressing regulatory requirements, gaining market differentiation, and identifying opportunities for innovation and

Objective:

The objective of this case study is to illustrate how Company Y leveraged rapid prototyping methods to expedite the design, testing, and refinement of pharmaceutical products.

Methodology:

Company Y collaborated with a multidisciplinary team comprising pharmaceutical scientists, engineers, and design experts. They utilized FDM technology to rapidly prototype various drug delivery devices, including inhalers, transdermal patches, and implantable devices. The iterative design process involved conceptualizing ideas, creating digital models, and 3D printing prototypes for functional testing and evaluation. Feedback from stakeholders, including healthcare professionals and end-users, was incorporated into subsequent iterations to optimize device performance and usability.

Results:

Reduced time-to-market: By employing rapid prototyping techniques, Company Y significantly reduced the time required to develop and commercialize new drug delivery devices. Iterative design cycles were completed in weeks rather than months, enabling faster product launches and market entry.

Enhanced design iteration: FDM technology facilitated rapid iteration of device designs, allowing Company Y to explore multiple concepts and identify optimal configurations efficiently. Real-time feedback from stakeholders enabled iterative improvements, resulting in enhanced device functionality and user experience.

Cost savings: Traditional prototyping methods are often time-consuming and expensive. By utilizing FDM for rapid prototyping, Company Y minimized material waste and reduced prototyping costs, resulting in substantial cost savings throughout the product development lifecycle.

Improved stakeholder engagement: The iterative prototyping process fostered collaboration and engagement among stakeholders, including healthcare professionals, patients, and regulatory agencies. By involving end-users early in the design process, Company Y ensured that the final products met their needs and expectations.

Conclusion:

Company Y's successful implementation of rapid prototyping techniques, including FDM, demonstrates the value of agile product development methodologies in the pharmaceutical industry. By leveraging rapid prototyping, Company Y accelerated innovation, reduced time-to-market, and enhanced stakeholder engagement. This case study underscores the importance of embracing emerging technologies to drive efficiency and competitiveness in pharmaceutical product development.

9.4.3 TITLE: ON-DEMAND MANUFACTURING REVOLUTIONIZES PHARMACEUTICAL SUPPLY CHAINS: A CASE STUDY [22, 25–26]

Introduction:

Traditional manufacturing processes in the pharmaceutical industry often involve large-scale production runs and extensive inventory management. This case study explores how Company Z implemented on-demand manufacturing utilizing FDM to enhance supply chain flexibility and efficiency.

Objective:

The objective of this case study is to showcase how Company Z leveraged on-demand manufacturing to reduce inventory costs, minimize lead times, and improve responsiveness to changing market demands.

Methodology:

Company Z adopted a just-in-time (JIT) manufacturing approach, leveraging FDM technology for on-demand production of pharmaceutical components and devices. By maintaining digital inventories of 3D printable files, they could rapidly manufacture products in response to customer orders or fluctuations in demand. FDM-enabled customization allowed Company Z to offer personalized solutions while minimizing excess inventory and waste.

Results:

Reduced inventory costs: By transitioning to on-demand manufacturing, Company Z significantly reduced inventory carrying costs associated with storing finished goods and raw materials. Digital inventories of 3D printable files eliminated the need for large warehouses and excess inventory, leading to substantial cost savings.

Minimized lead times: Traditional manufacturing processes often involve long lead times from order placement to delivery. With on-demand manufacturing, Company Z reduced lead times by producing products only when needed, thereby enhancing responsiveness to customer orders and market demands.

Enhanced customization: FDM technology enabled Company Z to offer customized pharmaceutical products tailored to individual patient needs. By leveraging on-demand manufacturing, they could quickly modify designs and produce personalized dosage forms, medical devices, and implants, enhancing patient outcomes and satisfaction.

Improved sustainability: On-demand manufacturing reduced material waste and energy consumption compared to traditional mass production methods. By producing only what is needed, Company Z minimized environmental impact while maintaining operational efficiency.

Conclusion:

Company Z's successful implementation of on-demand manufacturing using FDM technology demonstrates the transformative potential of agile manufacturing strategies in the pharmaceutical industry. By embracing on-demand production, Company Z achieved cost savings, minimized lead times, enhanced customization, and improved sustainability. This case study underscores the importance of adopting flexible and responsive manufacturing approaches to meet evolving market demands and drive competitiveness in the pharmaceutical sector.

9.4.4 TITLE: ADVANCING TISSUE ENGINEERING WITH 3D PRINTING: A CASE STUDY IN PHARMACEUTICAL REGENERATIVE MEDICINE [18, 25–27]

Introduction:

Tissue engineering holds great promise for regenerative medicine, offering solutions for tissue repair, replacement, and regeneration. This case study explores how Company A utilized FDM to advance tissue engineering applications in the pharmaceutical industry.

Objective:

The objective of this case study is to illustrate how Company A leveraged 3D printing technology to fabricate tissue scaffolds, implants, and drug delivery systems for regenerative medicine applications.

Methodology:

Company A collaborated with academic researchers and clinicians to develop 3D printed constructs for tissue regeneration. They utilized FDM to fabricate biocompatible scaffolds with precise geometries, porous structures, and controlled release capabilities. The scaffolds were designed to mimic the extracellular matrix and provide a supportive environment for cell growth, proliferation, and differentiation. Company A conducted *in vitro* and *in vivo* studies to evaluate the efficacy and safety of the 3D printed constructs for various tissue engineering applications, including bone regeneration, cartilage repair, and wound healing.

Results:

Bone regeneration: Company A successfully 3D printed porous scaffolds using FDM, which supported the attachment and proliferation of osteoblasts. In animal studies, the scaffolds promoted new bone formation and integration with surrounding tissue, demonstrating their potential for bone regeneration applications.

Cartilage repair: 3D printed scaffolds developed by Company A facilitated the encapsulation and differentiation of chondrocytes, leading to the formation of cartilage-like tissue *in vitro*. In preclinical models of cartilage defects, the scaffolds supported tissue repair and regeneration, highlighting their suitability for cartilage repair applications.

Wound healing: Company A's 3D printed scaffolds incorporated growth factors and bioactive molecules to promote wound healing and tissue regeneration. In animal models of chronic wounds, the scaffolds accelerated wound closure, reduced inflammation, and stimulated tissue regeneration, offering a promising approach for wound care applications.

Drug delivery systems: In addition to scaffolds, Company A developed 3D printed drug delivery systems for localized and controlled release of therapeutic agents. By encapsulating drugs within biocompatible filaments, they achieved sustained release profiles tailored to specific tissue regeneration processes, enhancing therapeutic efficacy and minimizing systemic side effects.

Conclusion:

Company A's successful implementation of FDM for tissue engineering applications underscores the transformative potential of 3D printing technology in pharmaceutical regenerative medicine. By leveraging FDM, Company A advanced the development of tissue scaffolds, implants, and drug delivery systems for bone regeneration, cartilage repair, wound healing, and beyond. This case study highlights the importance of interdisciplinary collaboration and innovation in driving progress toward personalized regenerative therapies in the pharmaceutical industry.

9.4.5 TITLE: TRANSFORMING PHARMACEUTICAL MANUFACTURING: CONTINUOUS PRODUCTION WITH FDM [28–30]

Introduction:

Continuous manufacturing represents a paradigm shift in pharmaceutical production, offering advantages such as improved efficiency, reduced waste, and enhanced quality control. This case study explores how pharmaceutical companies have integrated FDM into continuous manufacturing processes, focusing on two key applications: oral solid dosage forms and transdermal patches.

Objective:

The objective of this case study is to illustrate how pharmaceutical companies have harnessed FDM technology to streamline continuous manufacturing processes, optimizing the production of oral solid dosage forms and transdermal patches.

Methodology:

Pharmaceutical companies collaborated with manufacturing experts and regulatory agencies to develop FDM-based continuous manufacturing platforms. For oral solid dosage forms, they formulated polymer-based filaments containing active pharmaceutical ingredients (APIs) and excipients suitable for FDM processing. The continuous extrusion and printing process allowed for precise control of dosage uniformity, tablet geometry, and drug release kinetics. Similarly, for transdermal patches, companies engineered FDM-printable matrices loaded with APIs and permeation enhancers. The continuous deposition and curing process enabled the fabrication of thin, flexible patches with controlled drug delivery profiles.

Results:

Continuous production of oral solid dosage forms: Company X implemented FDM-based continuous manufacturing for the production of oral solid dosage forms. By extruding polymer-based filaments containing APIs and excipients, they achieved precise dosage control and uniform drug distribution within tablets. Real-time monitoring and process control systems ensured consistent tablet quality and compliance with regulatory requirements. The continuous manufacturing approach resulted in shorter production cycles, reduced material waste, and enhanced scalability compared to traditional batch processing methods.

Continuous fabrication of transdermal patches: Company Y pioneered FDM-based continuous manufacturing for transdermal patches. By depositing and curing FDM-printable matrices loaded with APIs and permeation enhancers, they produced thin, flexible patches with controlled drug release profiles. Inline quality assurance systems monitored patch thickness, drug content, and adhesion properties, ensuring product uniformity and reliability. The continuous manufacturing platform enabled rapid scale-up and customization of patch designs, facilitating the development of patient-specific formulations and dosage strengths.

9.5 CONCLUSION

Pharmaceutical companies have demonstrated the viability of FDM technology in continuous manufacturing processes for oral solid dosage forms and transdermal patches. By leveraging FDM-based approaches, companies have optimized production efficiency, minimized waste, and improved product quality in compliance with regulatory standards. This chapter highlights the transformative impact of continuous manufacturing and FDM technology on pharmaceutical production, paving the way for agile, cost-effective, and patient-centric manufacturing solutions.

KEYWORDS

- **biocompatible filaments**
- **customized drug delivery system**
- **drug release kinetics**
- **oral solid dosage form**
- **rapid prototyping**
- **tissue regeneration**
- **transdermal patches**

REFERENCES

1. Sudin, M., Shamsudin, S., & Abdullah, M, A, (2024). Effect of part features on dimensional accuracy of fdm model. https://www.researchgate.net/publication/305265273_Effect_of_part_features_on_dimensional_accuracy_of_fdm_model (accessed May 14, 2024).
2. Jain, P., & Kuthe, A. M. (2013). Feasibility study of manufacturing using rapid prototyping: FDM approach. *Procedia Engineering, 63*, 4–11. https://doi.org/10.1016/j.proeng.2013.08.275.
3. Harun, W. S. W., Safian, S., & Idris, M. H. (2009). Evaluation of ABS patterns produced from FDM for investment casting process. *WIT Transactions on Engineering Sciences, 64*. https://doi.org/10.2495/MC090301.
4. Ingole, D. S., Kuthe, A. M., Thakare, S. B., & Talankar, A. S. (2009). Rapid prototyping—A technology transfer approach for development of rapid tooling. *Rapid Prototyping Journal, 15*(4), 280–290. https://doi.org/10.1108/13552540910979794/FULL/PDF.
5. Rapid ProtoGeometric Errors in CNC Machine Tools. Manufacturing and industrial engineering.typing—A technology transfer approach for development of rapid tooling—Google Search. https://www.google.com/search?q=Rapid+ProtoGeometric Errors in CNC Machine Tools. Manufacturing and industrial engineering.typing+–+A+technology+transfer+approach+for+development+of+rapid+tooling.&rlz=1C1ONGR_enIN981IN981&oq=Rapid+Prototyping+–+A+technology+transfer+approach+for+development+of+rapid+tooling.&gs_lcrp=EgZjaHJvbWUqBggAEEUYOzIGCAAQRRg70gEHMjkxajBqNKgCALACAQ&sourceid=chrome&ie=UTF-8 (accessed May 14, 2024).

6. Lennings, L. (1970). Selecting either layered manufacturing or CNC machining to build your prototype.
7. Agarwala, M. K., Van Weeren, R., Bandyopadhyayl, A., Whalen, P. J., Safari, A., & Danforth, S. C. (1996). Fused deposition of ceramics and metals : An overview.
8. Alexander, P., Allen, S., & Dutta, D. (1998). Part orientation and build cost determination in layered manufacturing. *CAD: Computer Aided Design, 30*(5), 343–356. https://doi.org/10.1016/S0010-4485(97)00083-3.
9. Ahn, D., Kweon, J. H., Kwon, S., Song, J., & Lee, S. (2009). Representation of surface roughness in fused deposition modeling. *Journal of Materials Processing Technology, 209*(15–16), 5593–5600. https://doi.org/10.1016/J.JMATPROTEC.2009.05.016.
10. Dey, A., & Yodo, N. (2019). A systematic survey of FDM process parameter optimization and their influence on part characteristics. *Journal of Manufacturing and Materials Processing, 3*(3). https://doi.org/10.3390/JMMP3030064.
11. Maji, P. K., Banerjee, P. S., & Sinha, A. (2008). Application of rapid prototyping and rapid tooling for development of patient-specific craniofacial implant: An investigative study. *International Journal of Advanced Manufacturing Technology, 36*(5–6), 510–515. https://doi.org/10.1007/S00170-006-0868-9.
12. Galetto, M., Verna, E., & Genta, G. (2021). Effect of process parameters on parts quality and process efficiency of fused deposition modeling. *Computers and Industrial Engineering, 156*. https://doi.org/10.1016/j.cie.2021.107238.
13. Boschetto, A., Giordano, V., & Veniali, F. (2012). Modelling micro geometrical profiles in fused deposition process. *International Journal of Advanced Manufacturing Technology, 61*(9–12), 945–956. https://doi.org/10.1007/S00170-011-3744-1.
14. Ahn, S. H., Lee, C. S., & Jeong, W. (2004). Development of translucent FDM parts by post-processing. *Rapid Prototyping Journal, 10*(4), 218–224. https://doi.org/10.1108/13552540410551333/FULL/PDF.
15. Beeley, P. R., & Smart, R. F. (2023). Investment casting. *Investment Casting*. https://doi.org/10.1201/9781003419228/INVESTMENT-CASTING-PHILIP-BEELEY-ROBERT-SMART.
16. Tiwary, V. K., Padmakumar, A., & Malik, V. (2022). Adhesive bonding of similar/dissimilar three-dimensional printed parts (ABS/PLA) considering joint design, surface treatments, and adhesive types. *Proceedings of the Institution of Mechanical Engineers, Part C: Journal of Mechanical Engineering Science, 236*(16), 8991–9002. https://doi.org/10.1177/09544062221089849.
17. Anitha, R., Arunachalam, S., & Radhakrishnan, P. (2001). Critical parameters influencing the quality of prototypes in fused deposition modelling. *Journal of Materials Processing Technology, 118*(1–3), 385–388. https://doi.org/10.1016/S0924-0136(01)00980-3.
18. Armillotta, A. (2006). Assessment of surface quality on textured FDM prototypes. *Rapid Prototyping Journal, 12*(1), 35–41. https://doi.org/10.1108/13552540610637255.
19. Bakar, N. S. A., Alkahari, M. R., & Boejang, H. (2010). Analysis on fused deposition modelling performance. *Journal of Zhejiang University Science A, 11*(12), 972–977. https://doi.org/10.1631/JZUS.A1001365/METRICS.
20. Boschetto, A., Giordano, V., & Veniali, F. (2013). 3D roughness profile model in fused deposition modelling. *Rapid Prototyping Journal, 19*(4), 240–252. https://doi.org/10.1108/13552541311323254.

21. (PDF) Impact of the fused deposition (FDM) printing process on polylactic acid (PLA) chemistry and structure. https://www.researchgate.net/publication/317377066_Impact_of_the_Fused_Deposition_FDM_Printing_Process_on_Polylactic_Acid_PLA_Chemistry_and_Structure (accessed May 14, 2024).
22. Valino, A. D., Dizon, J. R. C., Espera, A. H., Chen, Q., Messman, J., & Advincula, R. C. (2019). Advances in 3D printing of thermoplastic polymer composites and nanocomposites. *Progress in Polymer Science, 98*, 101162. https://doi.org/10.1016/J.PROGPOLYMSCI.2019.101162.
23. Arenas, J. M., Alía, C., Blaya, F., & Sanz, A. (2012). Multi-criteria selection of structural adhesives to bond ABS parts obtained by rapid prototyping. *International Journal of Adhesion and Adhesives, 33*, 67–74. https://doi.org/10.1016/j.ijadhadh.2011.11.005.
24. Bernard, A., Delplace, J. C., Perry, N., & Gabriel, S. (2003). Integration of CAD and rapid manufacturing for sand casting optimisation. *Rapid Prototyping Journal, 9*(5), 327–333. https://doi.org/10.1108/13552540310502220/FULL/HTML.
25. Cheah, C. M., Chua, C. K., Lee, C. W., Feng, C., & Totong, K. (2005). Rapid prototyping and tooling techniques: A review of applications for rapid investment casting. *International Journal of Advanced Manufacturing Technology, 25*(3–4), 308–320. https://doi.org/10.1007/S00170-003-1840-6.
26. Lee, W. C., Wei, C. C., & Chung, S. C. (2014). Development of a hybrid rapid prototyping system using low-cost fused deposition modeling and five-axis machining. *Journal of Materials Processing Technology, 214*(11), 2366–2374. https://doi.org/10.1016/J.JMATPROTEC.2014.05.004.
27. A review on various approach for process parameter optimization of fused deposition modeling (FDM) process and taguchi approach for optimization. https://www.researchgate.net/publication/266416206_A_review_on_various_approach_for_process_parameter_optimization_of_fused_deposition_modeling_FDM_process_and_taguchi_approach_for_optimization (accessed May 14, 2024).
28. Xie, J., Yang, Z., Wang, X., & Lai, X. (2018). A cloud service platform for the seamless integration of digital design and rapid prototyping manufacturing. *International Journal of Advanced Manufacturing Technology, 100*(5), 1475–1490. https://doi.org/10.1007/S00170-018-2809-9.
29. Lin, S., Xia, L., Ma, G., Zhou, S., & Xie, Y. M. (2019). A Maze-like path generation scheme for fused deposition modeling. *International Journal of Advanced Manufacturing Technology, 104*(1–4), 1509–1519. https://doi.org/10.1007/S00170-019-03986-7/METRICS.
30. Galetto, M., Verna, E., & Genta, G. (2021). Effect of process parameters on parts quality and process efficiency of fused deposition modeling. *Computers and Industrial Engineering, 156*. https://doi.org/10.1016/J.CIE.2021.107238.

CHAPTER 10

Future Trajectories in Pharmaceutical Fused Deposition Modeling

LALITA NEMADE[1*], KIRTIMALA FALAK[2], MRUNAL KARANJKAR[3], MITALI PATIL[2], and SHEETAL MANE[4]

[1]*Department of Pharmaceutics, Govindrao Nikam College of Pharmacy, Sawarde, Maharashtra, India*

[2]*Department of Pharmaceutics, B. K. Patil College of Pharmacy, Taloja, Navi Mumbai, Maharashtra, India*

[3]*Department of Pharmacology, Govindrao Nikam College of Pharmacy, Sawarde, Maharashtra, India*

[4]*School of Pharmacy, Devi Ahilya Vishwavidyalaya, Indore, Madhya Pradesh, India; NMT Gujarati College of Pharmacy, Indore, Madhya Pradesh, India*

*Corresponding author

ABSTRACT

Pharmaceutical fused deposition modeling (FDM) is at the forefront of innovation in drug development and personalized medicine, offering transformative capabilities for the fabrication of customized dosage forms and advanced drug delivery systems. This chapter explores the evolving landscape and future directions of pharmaceutical FDM, highlighting critical advancements in materials science, including biocompatible, bioresorbable, and stimuli-responsive polymers. It examines innovations in precision printing technologies such as high-resolution systems, multi-material and multi-drug printing, and integration of in-line monitoring for quality control. The discussion also addresses key regulatory challenges

Advancements in Fused Deposition Modeling (FDM) Technology for Pharmaceutical and Medical Applications. Kuldeep Vinchurkar, Dinesh Kumar Mishra, Pankaj V Dixit, and Sheetal Mane (Eds.)

DOI: 10.1201/9781779640338-10

These polymers are capable of meeting the safety and efficacy requirements for drug delivery applications. Examples include PLA (polylactic acid), PCL (polycaprolactone), and PLGA (poly(lactic-co-glycolic acid)). Drug-loaded filaments: Researchers have been successful in incorporating APIs into FDM filaments. This allows for the direct 3D printing of drug-loaded dosage forms with precise control over drug dosage and release kinetics. Multi-material printing: Advancements in FDM technology have enabled the printing of multi-material structures, allowing for the fabrication of complex drug delivery systems with varying properties in different regions of the dosage form [5–7].

10.2.1 PROCESS OPTIMIZATION

1. **Printability and resolution:** Efforts have been made to optimize FDM printing parameters such as nozzle diameter, layer height, and printing speed to improve print quality, resolution, and reproducibility of pharmaceutical dosage forms.
2. **Post-processing techniques:** Post-processing techniques such as thermal treatment, solvent vapor smoothing, and surface coating have been explored to further enhance the mechanical properties, drug release profiles, and patient acceptability of FDM-printed dosage forms [8].
3. **Scale-up and manufacturing integration:** Scaling up FDM-based pharmaceutical manufacturing processes while maintaining product quality and regulatory compliance remains a challenge. Integration of FDM into traditional pharmaceutical manufacturing workflows requires careful consideration of GMP (good manufacturing practice) guidelines and process validation [5, 9–11].

10.2.2 REGULATORY CONSIDERATIONS

1. **Quality control and validation:** Regulatory agencies such as the FDA (Food and Drug Administration) require rigorous quality control and validation of 3D-printed pharmaceutical products. Manufacturers must demonstrate consistency, safety, and efficacy through comprehensive testing and documentation.

2. **Material safety and biocompatibility:** Ensuring the safety and biocompatibility of FDM-printed dosage forms and materials is essential. Manufacturers need to thoroughly evaluate the chemical composition, leachables, and extractables of printing materials to mitigate potential risks to patient health.
3. **Regulatory pathways:** Regulatory pathways for the approval of 3D-printed pharmaceuticals are still evolving. Collaborative efforts between industry stakeholders and regulatory agencies are necessary to establish clear guidelines and standards for the assessment and approval of FDM-printed dosage forms.

In summary, pharmaceutical FDM holds great promise for personalized medicine, on-demand manufacturing, and the development of novel drug delivery systems. However, addressing challenges related to materials science, process optimization, and regulatory considerations is crucial for realizing its full potential in the pharmaceutical industry [10, 11].

10.2.3 ADVANCES IN MATERIALS SCIENCE FOR PFDM

Materials science plays a pivotal role in the evolution of additive manufacturing techniques like precision fused deposition modeling (pFDM), enabling the fabrication of intricate structures with enhanced functionalities [5, 12, 13].

10.2.3.1 BIOCOMPATIBLE AND PHARMACOLOGICALLY RELEVANT MATERIALS

One of the foremost areas of innovation in pFDM materials science involves the synthesis and utilization of biocompatible materials with pharmacological relevance. These materials are essential for various applications ranging from tissue engineering to drug delivery systems. Recent developments have seen the emergence of biocompatible polymers such as PLA, polyglycolic acid (PGA), and their copolymers. These polymers exhibit excellent biocompatibility, allowing for their safe incorporation into medical devices and implants. Moreover, the integration of pharmacologically active compounds within these materials opens avenues for targeted drug delivery, enabling precise and sustained release kinetics. Functionalized polymers, including those loaded with growth factors or antimicrobial agents, demonstrate promise in promoting tissue regeneration and combating infections, thus revolutionizing biomedical applications of pFDM [4].

10.2.3.2 IMPROVED MECHANICAL AND CHEMICAL PROPERTIES

Enhancing the mechanical and chemical properties of materials used in pFDM is imperative for achieving superior performance and durability of fabricated components. Recent advancements have focused on optimizing polymer formulations and processing parameters to impart enhanced mechanical strength, toughness, and chemical resistance. Nanocomposite materials, reinforced with nanoparticles like carbon nanotubes or graphene, exhibit remarkable improvements in mechanical properties, including tensile strength and modulus. Furthermore, the incorporation of crosslinking agents or additives enhances chemical resistance, rendering the printed parts suitable for challenging environments such as automotive and aerospace applications. These advancements not only expand the scope of pFDM in industrial sectors but also pave the way for the production of functional prototypes with performance characteristics akin to traditional manufacturing methods [5, 10, 14].

10.2.3.3 BIORESORBABLE AND SUSTAINED-RELEASE FORMULATIONS

The development of bioresorbable materials and sustained-release formulations represents a significant breakthrough in pFDM, particularly in the realm of biomedical applications. Bioresorbable polymers, such as PLGA and PCL, degrade over time within the body, eliminating the need for implant removal surgeries. This property is instrumental in fabricating temporary medical implants, such as tissue scaffolds or drug-eluting stents, which gradually degrade as new tissue forms or drugs are released. Furthermore, the integration of controlled-release mechanisms within these materials enables precise modulation of drug release kinetics, offering prolonged therapeutic effects and minimizing systemic side effects. This capability holds immense potential for personalized medicine, where patient-specific drug dosages and release profiles can be tailored to optimize treatment outcomes. Moreover, the versatility of pFDM allows for the fabrication of intricate geometries with precisely controlled drug distribution, enhancing the efficacy of therapeutic interventions while minimizing dosage requirements. By enabling the fabrication of complex structures with tailored functionalities across diverse fields, new heights can be reached. Biocompatible and pharmacologically relevant materials have revolutionized biomedical applications, facilitating the development of implants and drug delivery systems with enhanced biocompatibility and therapeutic efficacy. Improved mechanical and chemical properties have expanded the industrial utility of pFDM, enabling the production of robust components for demanding applications. Moreover, the advent of bioresorbable materials and sustained-release formulations holds promise for personalized medicine, offering precise control over drug delivery and tissue regeneration. As material science continues to evolve, pFDM is poised to usher in a new era of innovation, driving advancements in healthcare, manufacturing, and beyond [2, 8, 15, 16].

10.3 INNOVATIONS IN PFDM EQUIPMENT AND TECHNOLOGY

pFDM has emerged as a versatile additive manufacturing technique, offering unprecedented flexibility in fabricating complex structures with intricate details. These advancements signify a paradigm shift in additive manufacturing, unlocking new possibilities across various industries [17].

10.3.1 HIGH-RESOLUTION PRINTING SYSTEMS

The quest for higher resolution in pFDM has led to significant advancements in printing systems, enabling the fabrication of intricate geometries with finer details and improved surface quality. Traditional pFDM systems often struggled with layer thickness limitations, resulting in visible layer lines and reduced surface finish. However, recent innovations have overcome these limitations through the integration of high-precision extrusion mechanisms, advanced nozzle designs, and improved motion control systems. Micro-extrusion techniques, utilizing ultrafine nozzles, and precise filament control, allow for layer thicknesses as low as tens of micrometers, drastically reducing the visibility of layer lines and enhancing part resolution. Furthermore, advancements in nozzle heating and cooling systems ensure precise filament deposition, minimizing droplet formation, and improving dimensional accuracy. These high-resolution printing systems empower designers and engineers to create intricate prototypes and functional parts with exceptional detail and fidelity, expanding the application scope of pFDM across industries such as aerospace, healthcare, and consumer electronics [17–19].

10.3.2 MULTI-MATERIAL AND MULTI-DRUG PRINTING CAPABILITIES

The ability to print with multiple materials or incorporate multiple drugs within a single print has emerged as a game-changer in pFDM technology, enabling the fabrication of complex, multifunctional structures with diverse properties and functionalities. Traditional pFDM systems were limited to printing with a single material, constraining the design possibilities and functionality of printed parts. However, recent innovations have introduced multi-material extrusion systems capable of seamlessly switching between different filaments during the printing process [3, 7]. This allows for the creation of composite structures with varying mechanical, thermal, or electrical properties, opening avenues for lightweight yet robust designs in automotive, aerospace, and sporting goods industries. Moreover, advancements in bio-printing technology have facilitated the integration of multiple drugs or bioactive agents within printed scaffolds or implants, enabling precise control over drug release kinetics and therapeutic efficacy. This capability holds immense potential in personalized medicine, where patient-specific drug formulations can be tailored to address individual healthcare needs. By enabling the fabrication of multifunctional structures with diverse material compositions, multi-material and multi-drug printing capabilities herald a new era of customization and innovation in additive manufacturing [7, 15].

10.3.3 INTEGRATION OF IN-LINE MONITORING AND QUALITY CONTROL

Ensuring the reliability and consistency of printed parts is paramount in additive manufacturing, particularly in industrial applications where stringent quality standards must be met. The integration of in-line monitoring and quality control mechanisms within pFDM systems has emerged as a key strategy to enhance process stability, detect defects in real-time, and optimize print quality. Advanced sensor technologies, including thermal imaging cameras, laser profilometers, and spectroscopic analyzers, enable real-time monitoring of key process parameters such as temperature, filament flow rate, and material properties. This allows for immediate detection of deviations from desired specifications, enabling prompt adjustments to maintain print quality and consistency. Furthermore, the integration of machine learning algorithms and artificial intelligence-based systems facilitates predictive maintenance and optimization of printing parameters based on historical data and performance trends. By leveraging in-line monitoring and quality control, pFDM systems can achieve higher yields, reduce waste, and improve overall process efficiency, making them increasingly viable for high-volume production environments in industries ranging from automotive manufacturing to medical device production [14]. High-resolution printing systems have revolutionized part quality, allowing for the creation of detailed prototypes and end-use parts with superior surface finish and dimensional accuracy. Multi-material and multi-drug printing capabilities have expanded the design possibilities and application scope of pFDM, empowering industries to create complex multifunctional structures tailored to specific requirements. The integration of in-line monitoring and quality control mechanisms ensures process stability, reliability, and consistency, paving the way for widespread adoption of pFDM in industrial manufacturing settings. As these advancements continue to evolve, pFDM is poised to revolutionize manufacturing across diverse sectors, driving innovation, customization, and efficiency in the production of next-generation products and technologies [2, 20].

10.4 REGULATORY CONSIDERATIONS AND STANDARDS

As pFDM continues to gain prominence in various industries, including healthcare, aerospace, and automotive, navigating the regulatory landscape is paramount to ensure compliance, safety, and efficacy of printed products. Understanding these considerations is essential for manufacturers, regulators, and stakeholders to foster innovation while upholding quality and safety standards [5, 16].

10.4.1 CURRENT REGULATORY LANDSCAPE FOR PFDM

The regulatory landscape for pFDM encompasses a complex web of regulations, standards, and guidelines established by governmental agencies and industry organizations worldwide. In sectors such as healthcare, where pFDM is utilized for medical device fabrication, regulatory oversight is stringent to safeguard patient health and ensure product performance. Regulatory bodies such as the FDA in the United States, the European Medicines Agency (EMA) in Europe, and the Pharmaceuticals and Medical Devices Agency (PMDA) in Japan set forth regulations governing the design, manufacturing, and marketing of medical devices fabricated using pFDM. These regulations require manufacturers to adhere to stringent quality management systems, conduct thorough risk assessments, and provide comprehensive documentation to demonstrate product safety and efficacy. Similarly, in sectors like aerospace and automotive, where pFDM is employed for prototyping, tooling, and part production, adherence to industry-specific standards and regulations is essential to ensure component performance, reliability, and regulatory compliance. While the regulatory landscape for pFDM is diverse and dynamic, manufacturers must navigate these complexities diligently to bring innovative products to market while mitigating regulatory risks [4, 21].

10.4.2 CHALLENGES AND OPPORTUNITIES IN REGULATORY APPROVAL

Navigating the regulatory approval process for pFDM possesses several challenges and opportunities for manufacturers and regulatory agencies alike. One of the primary challenges is the dynamic nature of additive manufacturing technologies, which often outpace the development of regulatory frameworks. Traditional regulatory pathways may not adequately address the unique considerations associated with pFDM, such as material variability, process validation, and post-processing requirements. Moreover, ensuring consistent quality and performance of printed products presents challenges due to the inherent variability in additive manufacturing processes [13]. Addressing these challenges requires collaboration between industry stakeholders and regulatory agencies to develop robust regulatory frameworks that balance innovation with safety and efficacy.

Despite these challenges, the regulatory approval process for pFDM also presents opportunities for manufacturers to demonstrate the benefits of additive manufacturing technologies, such as design flexibility, rapid prototyping, and customization. By proactively engaging with regulators, manufacturers can influence the development of regulatory standards and guidelines that accommodate the unique characteristics of pFDM while ensuring product quality and patient safety [9, 13].

10.4.3 EMERGING STANDARDS FOR MATERIAL CHARACTERIZATION AND PRINTING PROCESSES

The emergence of standardized protocols for material characterization and printing processes is instrumental in facilitating regulatory compliance and ensuring product quality in pFDM. Industry organizations such as ASTM International, ISO (International Organization for Standardization), and FDA have been instrumental in developing and updating standards specific to additive manufacturing technologies, including pFDM. These standards encompass a wide range of topics, including material properties, process parameters, post-processing techniques, and part performance evaluation. For example, ASTM F2792 outlines standard terminology, testing methods, and reporting requirements for additive manufacturing technologies, including pFDM, to facilitate communication and data exchange within the industry. Similarly, ISO standards such as ISO/ASTM 52900 provide guidelines for additive manufacturing terminology, while ISO 13485 specifies quality management system requirements for medical device manufacturers. By adhering to these standards, manufacturers can establish robust quality control processes, validate printing parameters, and ensure consistency and traceability throughout the product lifecycle. Furthermore, emerging standards for material characterization, such as ASTM F2971 for mechanical testing of additive manufactured parts, enable comprehensive evaluation of printed components' mechanical properties, aiding in regulatory submissions and product validation [6, 14, 22]. Navigating the complex regulatory landscape requires manufacturers to proactively engage with regulatory agencies, adhere to industry standards, and implement robust quality management systems to ensure compliance, safety, and efficacy of printed products. While challenges exist in aligning regulatory frameworks with the dynamic nature of additive manufacturing technologies, opportunities abound for manufacturers to influence regulatory developments, demonstrate the benefits of pFDM, and drive innovation in product design and development. By embracing emerging standards for material characterization and printing processes, manufacturers can establish best practices, streamline regulatory approval processes, and unlock the full potential of pFDM in revolutionizing manufacturing paradigms and delivering value-added solutions to global markets [2].

10.5 SCIENCE OF FDM AND APPLICATIONS

FDM, also known as fused filament fabrication (FFF), is a 3D printing technology that has gained widespread popularity due to its affordability, versatility, and ease of use. It involves the layer-by-layer deposition of thermoplastic materials to create three-dimensional objects. Developed by Scott Crump in the late 1980s, FDM has since evolved into a widely adopted additive manufacturing process utilized in various industries, including prototyping, manufacturing, education, and even healthcare [14, 23].

10.5.1 PRINCIPLE OF FDM

At the core of FDM lies a simple yet effective principle—the extrusion of molten thermoplastic material through a heated nozzle onto a build platform. The process begins with a spool of thermoplastic filament, typically made of materials such as ABS (acrylonitrile butadiene styrene) or PLA. This filament is fed into the printer's extrusion mechanism, where it undergoes heating and melting. The molten filament is then deposited layer by layer onto the build platform, with each layer fusing with the previous one as it cools down. This additive approach allows for the creation of complex geometries with relatively low material wastage [4].

10.5.1.1 EXTRUSION MECHANISM

The extrusion mechanism of an FDM printer comprises several key components, including the extruder assembly, heating element, filament feeder, and nozzle. The filament is driven into the extruder assembly by a motorized feeder, where it encounters a heating element that melts it to a precise temperature suitable for extrusion. The molten filament is then forced through a fine nozzle, whose diameter determines the width of the extruded material. The movement of the extruder assembly is controlled by computer-aided design (CAD) software, which dictates the deposition path and ensures precise positioning of the material.

10.5.1.2 LAYERING TECHNIQUES

FDM builds objects layer by layer, with each layer typically ranging from 0.1 to 0.3 mm in thickness. The layer height, also known as the layer resolution, plays a crucial role in determining the surface finish and structural integrity of the final print. Thinner layers result in smoother surfaces but require longer print times, while thicker layers may sacrifice surface quality for speed. Layer adhesion is paramount in FDM printing, as poor bonding between layers can lead to structural weaknesses and defects in the final object.

10.5.1.3 HEAT MANAGEMENT

Temperature control is essential in FDM printing to ensure proper melting and extrusion of the thermoplastic filament. The nozzle temperature must be precisely regulated to maintain the filament at its melting point without causing degradation or burning. Additionally, some FDM printers feature heated build platforms, which help prevent warping and improve adhesion between layers. The temperature settings may vary depending on the type of thermoplastic filament used, with different materials requiring different processing temperatures.

10.5.1.4 COOLING PROCESSES

As each layer of molten filament is deposited, it rapidly cools and solidifies to form the desired shape. Effective cooling is essential to prevent deformation, warping, and other printing defects. Many FDM printers are equipped with cooling fans or other cooling mechanisms positioned near the nozzle to expedite the cooling process. These fans help dissipate heat from the freshly deposited material, allowing it to solidify quickly and maintain its shape. Proper cooling also enhances overhang performance, enabling the printing of intricate geometries without the need for support structures [1, 5, 16].

10.5.2 ADVANCEMENTS AND FUTURE DIRECTIONS

FDM technology continues to evolve, driven by ongoing research and development efforts aimed at improving print quality, speed, and material compatibility. Recent advancements in FDM include the development of high-temperature thermoplastics, such as PEEK (polyether ether ketone) and ULTEM (polyetherimide), which offer enhanced mechanical properties and chemical resistance. Additionally, innovations in multi-material printing, automated support generation, and post-processing techniques are expanding the capabilities of FDM printers, opening up new possibilities for applications in various industries [3, 17, 22].

10.5.3 APPLICATIONS OF FDM

10.5.3.1 PROTOTYPING

One of the primary applications of FDM is rapid prototyping. In product development and design, FDM enables designers and engineers to quickly iterate on their concepts and test them in real-world scenarios. The speed and cost-effectiveness of FDM make it an ideal solution for producing functional prototypes, allowing for faster design validation and iteration cycles. Additionally, FDM offers the flexibility to create prototypes with complex geometries and intricate details, helping designers visualize their ideas more accurately.

10.5.3.2 CUSTOMIZATION

FDM empowers manufacturers to produce custom or low-volume parts without the need for expensive molds or tooling. This customization capability is particularly valuable in industries such as aerospace, automotive, and healthcare, where personalized components are often required to meet specific requirements. With FDM, manufacturers can easily modify designs to accommodate individual preferences, ergonomic considerations, or unique specifications, ultimately delivering tailored solutions to their customers.

10.5.3.3 EDUCATION

FDM printers have become invaluable tools in educational settings, providing students with hands-on learning experiences in design, engineering, and manufacturing. By using FDM technology, students can bring their creative ideas to life, design functional prototypes, and explore the principles of additive manufacturing firsthand. FDM printers are commonly found in schools, universities, and maker spaces, where they serve as catalysts for innovation and skill development among students of all ages.

10.5.3.4 MANUFACTURING AIDS

FDM is widely used to produce custom jigs, fixtures, and tooling for manufacturing processes. These aids play a crucial role in improving efficiency, accuracy, and safety on the factory floor. By leveraging FDM technology, manufacturers can rapidly prototype and iterate on jigs and fixtures to meet the evolving needs of their production lines. FDM-produced manufacturing aids are lightweight, durable, and cost-effective, offering significant advantages over traditional manufacturing methods [7].

10.5.4 ART AND DESIGN

Beyond industrial applications, FDM is also embraced by artists and designers as a medium for creative expression. From sculptures and architectural models to wearable art and home decor, FDM enables artists to materialize their visions with precision and efficiency. The accessibility of FDM technology allows artists to experiment with form, texture, and color, pushing the boundaries of traditional art and design. Moreover, FDM's sustainability and customization features align with contemporary artistic practices focused on environmental consciousness and personalization.

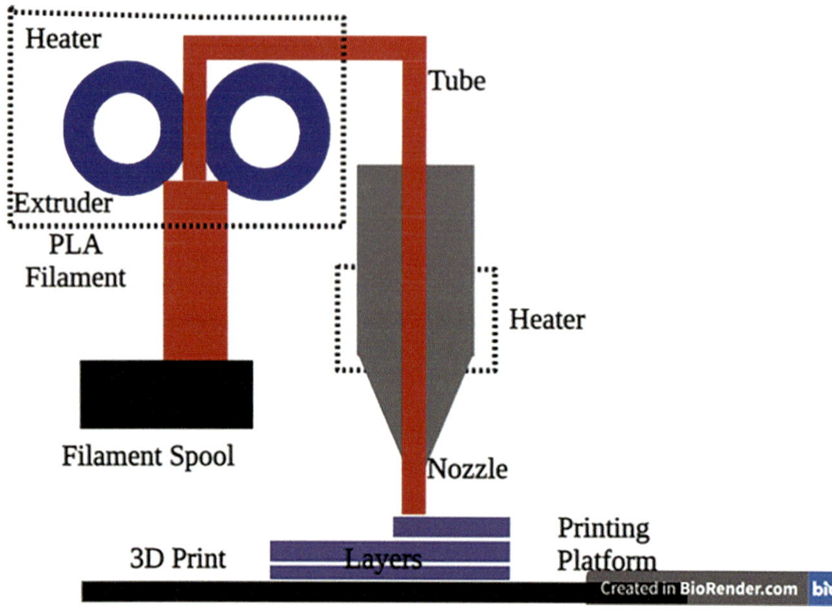

FIGURE 10.2 3D printing technique.

10.5.5 3D PRINTING AND THE NOVEL CORONAVIRUS (COVID-19) PANDEMIC

10.5.5.1 THE ROLE OF 3D PRINTING IN PPE PRODUCTION

During the COVID-19 pandemic, personal protective equipment (PPE) emerged as a critical defense against the virus, safeguarding frontline healthcare workers and the public. However, traditional manufacturing methods faced significant challenges in meeting the escalating demand for PPE. In response, 3D printing emerged as a pivotal solution, facilitating the rapid production of essential gear while minimizing setup time and costs.

3D printing technology empowered innovators worldwide to swiftly mobilize their resources, leveraging open-source designs and collaborating closely with healthcare professionals to ensure the efficacy and safety of the produced PPE. Face shields, masks, and respirators were among the key items manufactured using 3D printing. One of the notable advantages of 3D printing in PPE production was its flexibility. Unlike traditional manufacturing processes that require time-consuming retooling for each product iteration, 3D printers could seamlessly adapt to various designs and specifications, enabling quick adjustments based on evolving requirements or feedback from end-users. Moreover, the decentralized nature of 3D printing allowed for localized production, reducing supply chain vulnerabilities and ensuring timely delivery of PPE to areas in need. This distributed manufacturing model not only enhanced resilience in the face of global disruptions but also fostered community engagement and collaboration as individuals, businesses, and organizations joined forces to address the PPE shortage. Despite its effectiveness in addressing immediate needs, challenges such as material limitations, quality control, and scalability persisted within 3D printing-based PPE production. Nevertheless, the pandemic accelerated innovation in this field, spurring advancements in materials science, printing techniques, and regulatory frameworks to enhance the reliability and scalability of 3D-printed PPE [2–5, 13, 14].

10.5.5.2 VENTILATOR COMPONENTS AND MEDICAL DEVICES

Early in the COVID-19 pandemic, the shortage of ventilators, crucial for treating severe cases, posed a daunting challenge for healthcare systems worldwide. In response, 3D printing emerged as a vital technology for manufacturing ventilator components, including valves and connectors, offering a solution for rapid prototyping and production iteration. The versatility of 3D printing enabled researchers and engineers to swiftly design and produce ventilator parts, overcoming traditional manufacturing constraints and accelerating the development process. By leveraging 3D printing, they could create intricate components with precision, ensuring compatibility and functionality within complex ventilator systems. Furthermore, 3D printing facilitated the creation of low-cost, open-source ventilators tailored for emergency situations in resource-constrained settings. These initiatives exemplified the adaptability and accessibility of 3D printing technology in addressing urgent medical needs during the pandemic. Collaboration between healthcare professionals, engineers, and 3D printing experts played a crucial role in these efforts. By pooling their expertise and resources, they were able to rapidly iterate on designs, optimize functionality, and ensure compliance with regulatory standards [21, 24, 25]. The use of 3D printing for ventilator components not only helped mitigate supply shortages but also highlighted the potential for distributed

Looking ahead, collaborative efforts between academia, industry, and regulatory bodies will be instrumental in shaping the future trajectory of pharmaceutical FDM. Continued research into novel materials, printing techniques, and process optimization will drive innovation and address existing limitations.

In conclusion, the integration of FDM into pharmaceutical manufacturing represents a paradigm shift in drug development and delivery. As we navigate the complexities of this evolving field, it is imperative to remain vigilant, adaptable, and committed to advancing the frontiers of pharmaceutical science for the betterment of global healthcare.

KEYWORDS

- **biocompatible polymers**
- **bioresorbable polymers**
- **functional prototypes**
- **micro-extrusion techniques**
- **pharmaceutical dosage forms**
- **precision fused deposition modeling**
- **tissue scaffolds**

REFERENCES

1. Goyanes, A., Buanz, A. B. M., Basit, A. W., & Gaisford, S. (2014). Fused-filament 3D printing (3DP) for fabrication of tablets. *International Journal of Pharmaceutics, 476*(1), 88–92. https://doi.org/10.1016/j.ijpharm.2014.09.044.
2. Trenfield, S. J., Awad, A., Madla, C. M., Hatton, G. B., Firth, J., Goyanes, A., Gaisford, S., & Basit, A. W. (2019). Shaping the future: Recent advances of 3D printing in drug delivery and healthcare. *Expert Opinion on Drug Delivery, 16*(10), 1081–1094. https://doi.org/10.1080/17425247.2019.1660318.
3. Ventola, C. L. (2014). Medical applications for 3D printing: Current and projected uses. *P T. 39*(10), 704–11. .
4. Ilyés, K., Kovács, N. K., Balogh, A., Borbás, E., Farkas, B., Casian, T., Marosi, G., Tomuță, I., & Nagy, Z. K. (2019). The applicability of pharmaceutical polymeric blends for the fused deposition modelling (FDM) 3D technique: material considerations–printability–process modulation, with consecutive effects on in vitro release, stability and degradation. *European Journal of Pharmaceutical Sciences, 129*, 110–123. https://doi.org/10.1016/j.ejps.2018.12.019.

5. Norman, J., Madurawe, R. D., Moore, C. M. V., Khan, M. A., & Khairuzzaman, A. A New Chapter in Pharmaceutical Manufacturing: 3D-Printed Drug Products. *Advanced Drug Delivery Reviews, 108*, 39–50. https://doi.org/10.1016/j.addr.2016.03.001.
6. Alhnan, M. A., Okwuosa, T. C., Sadia, M., Wan, K. W., Ahmed, W., & Arafat, B. Emergence of 3D printed dosage forms: Opportunities and challenges. *Pharmaceutical Research, 33*(8), 1817–1832. https://doi.org/10.1007/S11095-016-1933-1.
7. Scoutaris, N., Ross, S., & Douroumis, D. (2016). Current trends on medical and pharmaceutical applications of inkjet printing technology. *Pharmaceutical Research, 33*(8), 1799–1816. https://doi.org/10.1007/s11095-016-1931-3.
8. Skowyra, J., Pietrzak, K., & Alhnan, M. A. Fabrication of Extended-Release Patient-Tailored Prednisolone Tablets via Fused Deposition Modelling (FDM) 3D Printing. *European Journal of Pharmaceutical Sciences, 68*, 11–17. https://doi.org/10.1016/j.ejps.2014.11.009.
9. Figueiredo, S., Fernandes, A. I., Carvalho, F. G., & Pinto, J. F. Performance and Paroxetine Stability in Tablets Manufactured by Fused Deposition Modelling-Based 3D Printing. *Journal of Pharmacy and Pharmacology, 74*(1), 67–76. https://doi.org/10.1093/JPP/RGAB138.
10. Tranová, T., Pyteraf, J., Kurek, M., Jamróz, W., Brniak, W., Spálovská, D., Loskot, J., Jurkiewicz, K., Grelska, J., Kramarczyk, D., et al. Fused Deposition Modeling as a Possible Approach for the Preparation of Orodispersible Tablets. *Pharmaceuticals, 15*(1). https://doi.org/10.3390/ph15010069.
11. Ononiwu, N., Ako, P., Anyaoha, C., Ikwuagwu, C., & Jacobs, I. (2024). Sustainable considerations in additive manufacturing processes: A review. *Journal of Mechanical Engineering Science, 18*(1), 9853–9871. https://doi.org/10.15282/jmes.18.1.2024.5.0780.
12. Zhang, J., Feng, X., Patil, H., Tiwari, R. V., & Repka, M. A. Coupling 3D printing with hot-melt extrusion to produce controlled-release tablets. *International Journal of Pharmaceutics, 519*(1–2), 186–197. https://doi.org/10.1016/j.ijpharm.2016.12.049.
13. Parulski, C., Jennotte, O., Lechanteur, A., & Evrard, B. (2021). Challenges of fused deposition modeling 3D printing in pharmaceutical applications: Where are we now? *Advanced Drug Delivery Reviews, 175*. https://doi.org/10.1016/j.addr.2021.05.020.
14. Feuerbach, T., Kock, S., & Thommes, M. (2018). *Characterisation of fused deposition modeling 3D printers for pharmaceutical and medical applications*, Vol. 23. Taylor & Francis. https://doi.org/10.1080/10837450.2018.1492618.
15. Tekin, E., Smith, P. J., & Schubert, U. S. Inkjet printing as a deposition and patterning tool for polymers and inorganic particles. *Soft Matter, 4*(4), 703–713. https://doi.org/10.1039/b711984d.
16. Jamróz, W., Szafraniec, J., Kurek, M., & Jachowicz, R. (2018). 3D printing in pharmaceutical and medical applications. *Pharmaceutical Research, 35*(9), Article 176.
17. Mahmood, M. A. (2021). 3D printing in drug delivery and biomedical applications: A state-of-the-art review. *Compounds, 1*(3), 94–115. https://doi.org/10.3390/compounds1030009.
18. Hong, M., Xu, L., Ren, G., Chen, J., & Qi, M. (2012). Fluid phase equilibria solubility of lansoprazole in different solvents. *Fluid Phase Equilibria, 331*, 18–25. https://doi.org/10.1016/j.fluid.2012.06.011.

19. Jyothirmai, P., Sumathi, S., Yuvraj, V., Anusha, A., & Suryakumari, Ch. (2018). A novel solubility enhancement method of Lansoprazole by mixed hydrotropy for the formulation and evaluation of fat-dissolving tablet of Lansoprazole. *European Journal of Biomedical and Pharmaceutical Sciences*, *5*(9), 284–293.
20. Chen-Wishart, M. (2010). *Understanding digital signal processing,* Third Edition. *Vascular,* No. January pp. 1–2.
21. Alhijjaj, M., Nasereddin, J., Belton, P., & Qi, S. (2019). Impact of processing parameters on the quality of pharmaceutical solid dosage forms produced by fused deposition modeling (FDM). *Pharmaceutics, 11*(12), 1–21. https://doi.org/10.3390/pharmaceutics11120633.
22. Chase, R. B., et al. (2021). Operations management for competitive advantage.
23. Basit, A. W., & Trenfield, S. J. (2022). 3D printing of pharmaceuticals and the role of pharmacy. *Pharmaceutical Journal, 308*(7959), 1–32. https://doi.org/10.1211/PJ.2022.1.135581.
24. Shi, K., Slavage, J. P., Maniruzzaman, M., & Nokhodchi, A. (2021). Role of release modifiers to modulate drug release from fused deposition modelling (FDM) 3D printed tablets. *International Journal of Pharmaceutics, 597*(January), 120315. https://doi.org/10.1016/j.ijpharm.2021.120315.
25. Wang, J., Goyanes, A., Gaisford, S., & Basit, A. W. (2016). Stereolithographic (SLA) 3D printing of oral modified-release dosage forms. *International Journal of Pharmaceutics, 503*(1–2), 207–212. https://doi.org/10.1016/j.ijpharm.2016.03.016.
26. Elkasabgy, N. A., Mahmoud, A. A., & Maged, A. (2020). 3D printing: An appealing route for customized drug delivery systems. *International Journal of Pharmaceutics, 588*. https://doi.org/10.1016/j.ijpharm.2020.119732.
27. Goole, J., & Amighi, K. (2016). 3D printing in pharmaceutics: A new tool for designing customized drug delivery systems. *International Journal of Pharmaceutics, 499*(1–2), 376–394. https://doi.org/10.1016/j.ijpharm.2015.12.071.

Index

L

Q

R

S

T

U

V